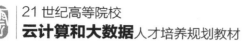

21世纪高等院校
云计算和大数据人才培养规划教材

云计算和大数据
技术实战

李俊杰 石慧 谢志明 谢高辉 唐华 王鹏 ◎ 编著

人 民 邮 电 出 版 社

北 京

图书在版编目（CIP）数据

云计算和大数据技术实战 / 李俊杰等编著. -- 北京：人民邮电出版社，2015.8（2021.6重印）
21世纪高等院校云计算和大数据人才培养规划教材
ISBN 978-7-115-39079-0

Ⅰ. ①云… Ⅱ. ①李… Ⅲ. ①计算机网络－数据处理－高等学校－教材 Ⅳ. ①TP393

中国版本图书馆CIP数据核字(2015)第078004号

内 容 提 要

本书是"云计算和大数据技术"课程的项目化实战教材，全书共 11 个项目，设计了 6 项主任务、34 项子任务。以搭建云计算平台流程组织实训内容，全面介绍了 CentOS 的安装及网络配置方法，虚拟化技术 KVM 的使用，MPI 的安装与部署，分布式处理系统 Hadoop 软件的编译、安装及部署，Hbase 数据库的安装与配置，以及大数据流处理系统 Storm 的安装与部署。本书突出上机操作，图文并茂且条理分明，实验丰富，把实验内容与课程教学相结合，有利于培养学生"做中学，学而会，会且熟"的学习能力。

本书可作为各级各类院校云计算、大数据、计算机相关专业课程的教材，也可作为 IT 类培训机构的云计算与大数据等相关课程的培训教程，还可供想从事云计算和大数据研究的读者自学使用。

◆ 编　著　李俊杰　石　慧　谢志明　谢高辉
　　　　　　唐　华　王　鹏
责任编辑　王　威
责任印制　杨林杰

◆ 人民邮电出版社出版发行　北京市丰台区成寿寺路 11 号
邮编 100164　电子邮件 315@ptpress.com.cn
网址 http://www.ptpress.com.cn
北京天宇星印刷厂印刷

◆ 开本：787×1092　1/16
印张：12.25　　　　　2015 年 8 月第 1 版
字数：321 千字　　　2021 年 6 月北京第 5 次印刷

定价：32.00 元

读者服务热线：(010)81055256　印装质量热线：(010)81055316
反盗版热线：(010)81055315

前言 PREFACE

随着计算机信息技术，特别是网络技术的发展，云计算与大数据技术的出现成为信息产业的重大变革。云计算技术将计算资源、存储资源以及相关各类广义的资源通过网络以服务的方式提供给资源的使用者，改变了传统信息技术架构中物理资源直接独占使用的模式。云计算、物联网、移动互联网的迅速发展催生了大数据时代的到来，从多种类型的数据中快速获得有价值信息的能力就是大数据技术，其核心价值是对海量数据的存储和分析。随着云计算相关产业的发展，社会需要大量懂技术、能应用的专业技术人才，"云计算技术与应用"专业已列入新版高等职业学校专业目录，云计算与大数据专业的人才培养面临全新的挑战和机遇。本书作为"十二五"职业教育国家规划教材《云计算与大数据技术》的配套实训教材具有以下特色和创新点。

面向企事业对应用型人才的需求特点，侧重对学生和读者操作技术能力的培养。本教材针对各级各类院校云计算与大数据专业应用型人才的培养目标、教学对象、教学特点和软硬件条件而编写，结合编者的一线教学经验，具有实用、操作性强等特点。作为云计算和大数据课程体系的一部分，本教材将使理论与实践教学更为密切、更为系统，以更好地满足实践教学要求。

本书以 11 个实战项目为主线，从基础入手，认真规划和组织教材内容，涵盖了 CentOS、CentOS 网络配置、虚拟化技术、MPI、Hadoop、HBase、Storm 等云计算与大数据技术的基础主流应用技术。每个实训项目又细化成若干个任务，环环相扣，前后衔接。每个任务均详细介绍了相关知识、任务内容和操作步骤，让读者在阅读过程中思路非常清楚。本书图文并茂、内容翔实、操作性强，每个任务都经过反复验证。我们把操作步骤中的大部分原始图片和操作命令保留在了教材中，使读者能准确掌握操作步骤；同时还注意引导学生主动学习、高效学习，让学生在完成任务后能提升操作技能、增长知识、学以致用。

本书在广东省高等教育学会高职高专云计算与大数据专业委员会理事长王鹏教授的组织和指导下编写完成，汕尾职业技术学院李俊杰老师编写项目三、六、七、八、九，石慧老师编写项目一、二、四、五，谢志明老师编写项目十、十一，广州五舟科技股份有限公司总经理谢高辉为本书提供云计算与大数据实训室建设解决方案，广州科技贸易职业学院王永祥副教授对本书的实验体系进行指导，五舟技术研究院唐华对本书编写进行技术指导，成都信息工程学院并行计算实训室刘峰、袁亚男对本书的实训进行验证。

本书是广东省高等教育学会高职高专云计算与大数据专业委员会2014年度教育科研立项课题（项目编号：GDYJSKT14-02、GDYJSKT14-09）、汕尾职业技术学院 2014 年度资源精品共享课《云计算技术》（项目编号：swzyjpkc14002）、广州市教育科学规划课题（项目编号：1201420456）和成都市科技局创新发展战略研究项目(项目编号:11RKYB016ZF)的科研成果，得到人民邮电出版社、广东省高职高专云计算与大数据专业委员会、广州五舟科技股份有限公司、汕尾市创新工业设计研究院的鼎力支持，同时也得到汕尾职业技术学院教务处、科研处、网络中心和数学与应用系领导、老师的支持和帮助，因为有了你们的支持和帮助，我们才能完成本书的出版工作。由于云计算与大数据技术涉及面广，在本书的编写过程中参考了大量学者、专家的资料，我们在这里向他们表示感谢。

为方便读者学习和满足教学需要，本实训教材配备了大量的电子资源，主要包括：项目资料、操作视频、软件资源等，欢迎读者登录到并行计算实验室网站 http://www.qhoa.org 免费下载使用，同时还欢迎相关课程的教师加入云计算与大数据教育 QQ 群(321168742)讨论交流。

我们在编写本书的过程中力求精益求精，但是由于编者经验和水平有限，书中的疏漏之处还敬请各位专家和广大读者批评指正。编者的 E-mail：gdswyun@126.com。

感谢您使用本书，期待本书能成为您的良师益友。

编者

2015 年 3 月

课程导学

高职院校教学特点在于"理论知识够用，重在实操实训"，云计算与大数据专业建设更是如此。由于该课程体系庞大、涉猎范围广，课程安排可采用周期反复性开设，由浅入深螺旋式上升来设置教学内容，帮助学生扎牢根基。在开展课程实训教学时，"练"字将为课堂教学的核心，熟能生巧，只有让学生动手操练多遍才能让他们牢牢地掌握这门课程、了解这个专业、爱上这个行业。通过本课程的学习，学生可以掌握云计算平台搭建的基本操作流程，掌握大数据的处理及管理模式，了解岗位所需的相应工作技能。

一、课程开设目标

本课程强调对学生职业岗位能力和职业素质的培养，在教学过程中针对不同环节采用恰当的教学方法。有意识、有步骤地将职业能力的培养融入实际教学过程中，最终使学生在课程学习的过程中逐渐养成良好的职业习惯和具备操作云计算的基本职业能力。

1. 方法能力目标

培养学生自学能力；培养学生勤于思考、做事认真的优良作风；培养学生良好的职业道德和吃苦耐劳、敬业乐业的工作作风。

2. 社会能力目标

培养学生的沟通能力和团队协作精神；培养学生分析问题、解决问题的能力；培养学生社会适应与应变的能力；培养学生接受新事物、新技术的能力；培养学生的商业头脑，使其具有质量、成本、生产安全等意识。

3. 岗位能力目标

掌握 Linux 的使用；掌握 KVM 技术的实现；掌握 MPI 的安装与运行；掌握 HBase 数据库管理；掌握 Hadoop、Storm 系统的安装、部署与使用。

4. 技能服务目标

可从事各企事业 IT 部门的工作；可从事大数据系统应用及开发的工作；可从事云计算平台搭建、测试、优化、管理和运维等日常工作。

二、课程内容组织安排

课程结合《云计算和大数据技术》教材组织教学，有理有据。按照云计算平台搭建的流程和学生认知规律精心设计了 11 个项目，6 项主任务、34 项子任务，学生在完成任务的同时，也就比较系统地掌握了云计算、大数据设计的基本思路和模式。

主任务	项目	典型子任务	知识点与技能点	学时
Linux 基本操作	1.搭建 CentOS 服务器	CentOS 安装	（1）CentOS 的分区及安装 （2）CentOS 用户及类型 （3）CentOS 登录界面的切换	4
Linux 基本操作	2.CentOS 网络管理	网络配置	（1）网络配置参数 （2）命令及图形界面下配置网络 （3）网络设备号及 MAC 地址 （4）安全设置	4
KVM 使用	3.虚拟化技术	（1）配置虚拟化网络 （2）虚拟系统管理器的使用 （3）虚拟机的运行与远程访问	（1）虚拟化网络的概念 （2）NAT 和 Bridge 的工作方式 （3）虚拟化概念 （4）虚拟机安装 （5）虚拟化系统管理器 virt-manager 的使用 （6）VNC 服务和 SPICE 服务的配置 （7）remote-viewer、VNC、SPICE 客户端工具的使用 （8）virsh-install 命令的使用 （9）virsh-close 命令克隆虚拟机 （10）virsh 常用命令及 qemu 命令的使用	10
MPI 部署	4.MPI——面向计算的集群技术	mpich 编译安装	（1）ssh 无密码访问配置 （2）NFS 服务器的基本概念 （3）NFS 服务器的配置方法及访问 （4）mpich 的编译与安装	4
MPI 部署	5.MPI 分布式程序设计基础	最简单的并行程序编写	（1）并行程序的开发思路及方法 （2）并行程序的编译与运行 （3）MPI 常用函数的使用 （4）并行程序的编写	4
Hadoop 部署	6.Hadoop 软件的编译打包	安装编译环境	（1）CentOS 编译命令 （2）ant 的作用 （3）编译环境的搭建 （4）maven 工具的使用	4
Hadoop 部署	7.Hadoop 环境的搭建与管理	Hadoop 的安装与配置	（1）hadoop 参数的设置 （2）hadoop 分布式环境的搭建 （3）使用浏览器监控 hadoop 服务状态 （4）hadoop shell 常用命令的使用	10
Hadoop 部署	8.Map/Reduce 实例	完成 Map/Reduce 项目	（1）Map/Reduce 的工作流程 （2）Map/Reduce 的单节点编程和运行 （3）Eclipse 工具的使用 （4）Map/Reduce 项目的编译打包及运行	6

续表

主任务	项目	典型子任务	知识点与技能点	学时
HBase 安装	9.HBase 分布式数据库	HBase 的安装与配置	（1）分布式数据库概念 （2）HBase 数据库的安装和配置 （3）HBase Shell 常用命令的使用 （4）安全设置	6
Storm 部署	10.Storm 环境的搭建与管理	Storm 的安装与配置	（1）Storm 的工作流程 （2）安装相关的依赖软件，如 JDK、Python 等 （3）安装运行 Storm 必备的工具包：ZeroMQ、JZMQ、Zookeeper （4）zookeeper 的配置 （5）Storm 的安装、环境搭建、管理和操作 （6）Storm 客户端常用操作命令	12
Storm 部署	11.Storm 拓扑实例	（1）使用 Eclipse 管理 Storm-Starter （2）编写拓扑实现单词计数	（1）Storm 基本术语 （2）拓扑的运行模式 （3）使用 mvn 对 Storm-Starter 项目进行打包 （4）Storm 拓扑的编程及运行 （5）pom 配置文件的编写 （6）使用 mvn 工程生成指定项目 （7）StormUI 管理	8

　　本课程建议高职高专类的院校在第三学期开设，计划 72 学时，前修课程主要有 Linux 的配置与管理、JAVA 基础、计算机网络管理，不同专业根据课程目标定位的不同可以适当调整课程内容和学时数。在教学实施过程中，为了保证教学效果和课程进度，可以事先将离线安装包提供给学生，以免课上学生同时进行在线编译或在下载时导致网络拥堵或瘫痪。

目 录 CONTENTS

第 1 章　搭建 CentOS 服务器　1

- 1.1 任务一　CentOS 安装　1
  - 1.1.1 任务描述　1
  - 1.1.2 相关知识　1
  - 1.1.3 任务实施　3
- 1.2 任务二　用户登录　11
  - 1.2.1 任务描述　11
  - 1.2.2 相关知识　12
  - 1.2.3 任务实施　12
- 1.3 本章小结　14

第 2 章　CentOS 网络管理　15

- 2.1 任务一　网络配置　15
  - 2.1.1 任务描述　15
  - 2.1.2 相关知识　15
  - 2.1.3 任务实施　16
- 2.2 任务二　更改以太网卡名称　23
  - 2.2.1 任务描述　23
  - 2.2.2 相关知识　23
  - 2.2.3 任务实施　24
- 2.3 任务三　关闭安全设置　25
  - 2.3.1 任务描述　25
  - 2.3.2 相关知识　25
  - 2.3.3 任务实施　25
- 2.4 本章小结　26

第 3 章　虚拟化技术　27

- 3.1 任务一　配置虚拟化网络　27
  - 3.1.1 任务描述　27
  - 3.1.2 相关知识　27
  - 3.1.3 任务实施　28
- 3.2 任务二　安装虚拟化软件包　31
  - 3.2.1 任务描述　31
  - 3.2.2 相关知识　31
  - 3.2.3 任务实施　32
- 3.3 任务三　虚拟系统管理器的使用　36
  - 3.3.1 任务描述　36
  - 3.3.2 相关知识　36
  - 3.3.3 任务实施　37
- 3.4 任务四　虚拟机的运行与远程访问　41
  - 3.4.1 任务描述　41
  - 3.4.2 相关知识　41
  - 3.4.3 任务实施　42
- 3.5 任务五　使用 virsh-install 安装虚拟机　48
  - 3.5.1 任务描述　48
  - 3.5.2 相关知识　48
  - 3.5.3 任务实施　49
- 3.6 任务六　使用 virsh-clone 克隆虚拟机　50
  - 3.6.1 任务描述　50
  - 3.6.2 相关知识　50
  - 3.6.3 任务实施　51
- 3.7 任务七　virsh 命令的使用　52
  - 3.7.1 任务描述　52
  - 3.7.2 相关知识　52
  - 3.7.3 任务实施　53
- 3.8 本章小结　55

第 4 章　MPI——面向计算的集群技术　56

- 4.1 任务一　配置 ssh 实现节点间无密码访问　56
  - 4.1.1 任务描述　56
  - 4.1.2 相关知识　56
  - 4.1.3 任务实施　58
- 4.2 任务二　网络文件系统 NFS　60
  - 4.2.1 任务描述　60
  - 4.2.2 相关知识　61
  - 4.2.3 任务实施　63
- 4.3 任务三　MPICH 编译运行　65
  - 4.3.1 任务描述　65
  - 4.3.2 相关知识　65
  - 4.3.3 任务实施　66
- 4.4 本章小结　71

第 5 章　MPI 分布式程序设计基础　72

- 5.1 任务一　最简单的并行程序的编号　72
  - 5.1.1 任务描述　72
  - 5.1.2 相关知识　72
  - 5.1.3 任务实施　74
- 5.2 任务二　获取进程标志和机器名的并行程序的编写　74
  - 5.2.1 任务描述　74
  - 5.2.2 相关知识　75
  - 5.2.3 任务实施　75
- 5.3 任务三　有消息传递功能的并行程序的编写　76
  - 5.3.1 任务描述　76
  - 5.3.2 相关知识　76
  - 5.3.3 任务实施　78
- 5.4 本章小结　79

第 6 章　Hadoop 软件的编译打包　80

- 6.1 任务一　安装编译环境　80
  - 6.1.1 任务描述　80
  - 6.1.2 相关知识　80
  - 6.1.3 任务实施　81
- 6.2 任务二　编译 Hadoop 软件　85
  - 6.2.1 任务描述　85
  - 6.2.2 相关知识　86
  - 6.2.3 任务实施　86
- 6.3 本章小结　87

项目 7　Hadoop 环境的搭建与管理　88

- 7.1 任务一　Hadoop 的安装与配置　88
  - 7.1.1 任务描述　88
  - 7.1.2 相关知识　88
  - 7.1.3 任务实施　90
- 7.2 任务二　Hadoop 的管理　95
  - 7.2.1 任务描述　95
  - 7.2.2 相关知识　95
  - 7.2.3 任务实施　95
- 7.3 任务三　Hadoop Shell 命令的使用　100
  - 7.3.1 任务描述　100
  - 7.3.2 相关知识　101
  - 7.3.3 任务实施　101
- 7.4 本章小结　104

第 8 章　Map/Reduce 实例　105

- 8.1　任务一　实现 Map/Reduce 的 C 语言实例　105
  - 8.1.1　任务描述　105
  - 8.1.2　相关知识　105
  - 8.1.3　任务实施　106
- 8.2　任务二　安装 Eclipse 开发工具　109
  - 8.2.1　任务描述　109
  - 8.2.2　相关知识　109
  - 8.2.3　任务实施　109
- 8.3　任务三　完成 Map/Reduce 项目　111
  - 8.3.1　任务描述　111
  - 8.3.2　相关知识　111
  - 8.3.3　任务实施　112
- 8.4　本章小结　121

第 9 章　HBase 分布式数据库　122

- 9.1　任务一　HBase 的安装与配置　122
  - 9.1.1　任务描述　122
  - 9.1.2　相关知识　122
  - 9.1.3　任务实施　123
- 9.2　任务二　HBase 管理与 HBase Shell 的使用　126
  - 9.2.1　任务描述　126
  - 9.2.2　相关知识　126
  - 9.2.3　任务实施　127
- 9.3　本章小结　132

第 10 章　Storm 环境的搭建与管理　133

- 10.1　任务一　Storm 的安装与配置　134
  - 10.1.1　任务描述　134
  - 10.1.2　相关知识　134
  - 10.1.3　任务实施　135
- 10.2　任务二　Storm 的管理　145
  - 10.2.1　任务描述　145
  - 10.2.2　相关知识　145
  - 10.2.3　任务实施　147
- 10.3　本章小结　151

第 11 章　Storm 拓扑实例　152

- 11.1　任务一　完成实例 Storm-Starter　152
  - 11.1.1　任务描述　152
  - 11.1.2　相关知识　152
  - 11.1.3　任务实施　153
- 11.2　任务二　使用 Eclipse 管理 Storm-Starter　159
  - 11.2.1　任务描述　159
  - 11.2.2　相关知识　159
  - 11.2.3　任务实施　160
- 11.3　任务三　编写拓扑实现单词计数　166
  - 11.3.1　任务描述　166
  - 11.3.2　相关知识　166
  - 11.3.3　任务实施　167
- 11.4　任务四　实现对文件单词计数　176
  - 11.4.1　任务描述　176
  - 11.4.2　任务实施　176
- 11.5　本章小结　186

第 1 章 搭建 CentOS 服务器

长期以来，Linux 一向是备受云计算和数据中心青睐的操作系统。云计算和大数据的存储通过 Internet 将物理资源（比如处理器和存储空间）转换成可伸缩的共享资源（将云计算和存储作为"服务"）。通过服务器虚拟化共享物理系统使得云计算和存储更加高效、伸缩性更强。通过云计算，用户可以访问大量的计算和存储资源，并且不必关心它们的位置和它们是如何配置的，Linux 在这个过程中扮演了重要的角色。

Linux 是一个多用户、多任务的操作系统，其运行方式、功能和 UNIX 系统很相似，但 Linux 系统的稳定性、安全性与网络功能是许多商业操作系统所无法比拟的。Linux 系统最大的特色是源代码完全公开，在符合 GNU/GPL（通用公共许可证）的原则下，任何人都可以自由取得、散布甚至修改源代码。

1.1 任务一 CentOS 安装

1.1.1 任务描述

CentOS 服务器的搭建是云计算与大数据的基础。本节任务是理解 Linux 系统特点，掌握 CentOS 的安装步骤，理解 Linux 的启动过程和运行级别。要想安装 CentOS 需要了解一下硬件需求，确认 CentOS 的分区参数，了解 CentOS 安装方式，最后完成 CentOS 的安装。

1.1.2 相关知识

1. 什么是 CentOS

CentOS（Community Enterprise Operating System，社区企业操作系统）是 Linux 发行版之一，它是 Red Hat Enterprise Linux 依照开放源代码规定释出的源代码所编译而成的，是安全、低维护、稳定、高预测性、高重复性的 Linux 环境。

2. CentOS 的特点

Linux 之所以备受云计算与大数据领域的青睐，与其自身的优良特性是分不开的。Linux 与其他操作系统相比，具有以下一系列显著的特点。

（1）源码公开，是自由软件。

Linux 可以说是作为开放源码的自由软件的代表，作为自由软件，它有两个特点：一是它开放源码并对外免费提供，二是爱好者可以按照自己的需要自由修改、复制和发布程序的源码。

（2）广泛的硬件支持，极强的平台可伸缩性。

Linux 能支持 x86、ARM、MIPS、ALPHA 和 PowerPC 等多种体系结构的微处理器。目前已成功地移植到数十种硬件平台，几乎能运行在所有流行的处理器上。Linux 能运行在笔记本电脑、台式计算机、工作站，直至巨型机上，而且几乎能在所有主要 CPU 芯片搭建的体系结构上运行（包括 Intel/AMD 及 HP-PA、MIPS、PowerPC、UltraSPARC、ALPHA 等 RISC 芯片），其伸缩性远远超过了 NT 操作系统目前所能达到的水平。

（3）安全性及可靠性好，内核高效稳定。

Linux 内核的高效和稳定已在各个领域内得到了大量事实的验证。Linux 中大量网络管理、网络服务等方面的功能，可使用户很方便地建立高效稳定的防火墙、路由器、工作站、服务器等。为提高安全性，它还提供了大量的网络管理软件、网络分析软件和网络安全软件等。

（4）真正的多任务多用户。

尽管许多操作系统声明支持多任务，但并不完全准确，如 Windows。而 Linux 则充分利用了 X86CPU 的任务切换机制，实现了真正多任务、多用户环境，允许多个用户同时执行不同的程序，并且可以给紧急任务以较高的优先级。

（5）具有强大的网络功能。

Linux 从诞生之日起就与 Internet 密不可分，支持各种标准的 Internet 网络协议，并且很容易移植到嵌入式系统中。目前，Linux 几乎支持所有主流的网络硬件、网络协议和文件系统，因此它是 NFS 的一个很好的平台。

（6）完全符合 POSIX 标准。

POSIX 是基于 UNIX 的第一个操作系统簇国际标准，Linux 遵循这一标准使 UNIX 下许多应用程序可以很容易地移植到 Linux 下，相反也是这样。

（7）具有丰富的图形用户界面。

Linux 的图形用户界面是 Xwindow 系统。Xwindow 可以做 MSWindows 下的所有事情，而且更有趣、更丰富，用户甚至可以在几种不同风格的窗口之间来回切换。

3．基本的硬件要求

安装 CentOS-6.5-x86_64 版的操作系统需要 CPU 支持 64 位，一般情况下，CPU 数量是双核或以上的普遍支持 64 位模式。另安装时如需启用图形界面模式，请保证安装机的内存最小不能低于 628MB。

4．UTC 时间

对于集群和分布式等大型系统来说，时间及其同步是个很重要的问题。通常的时间可分为 local 时间和 UTC 时间。

local 时间在 linux 下使用 date 查看，如 date。

UTC 时间是指标准格林威治及零时区的时间，不包含夏令时的计算。UTC 时间查看方式,如 date -u。

基本上,Local 时间=UTC 时间+时区时间差+夏令时间差。

还有硬件时间及 BIOS 上存储的时间,查看方式如 hwclock。

1.1.3 任务实施

准备工作

下载 CentOS-6.5-x86_64。CentOS 的官方网站为 http://www.centos.org/,镜像网站为 http://mirrors.yun-idc.com/centos/。

步骤 1

直接光盘安装,选择"Install or upgrade an existing system",如图 1.1 所示。

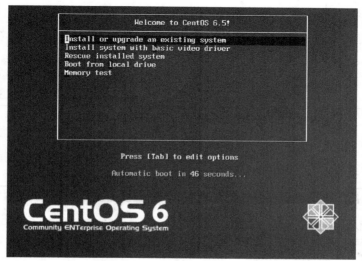

图 1.1 光盘安装选择"Install or upgrade an existing system"

步骤 2

不检查媒体,选择"Skip",如图 1.2 所示。

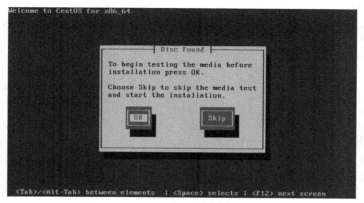

图 1.2 选择"Skip"

步骤 3

选择"Chinese(Simplified) (中文(简体))",如图 1.3 所示。

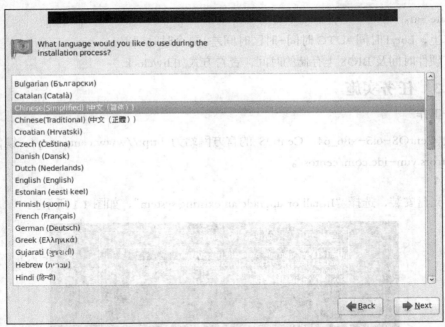

图 1.3 选择"Chinese(Simplified)(中文(简体))"

步骤 4

选择"美国英语式",如图 1.4 所示。

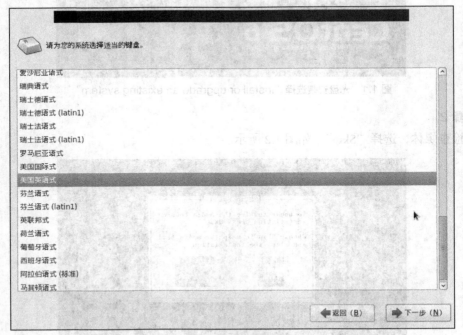

图 1.4 选择"美国英语式"

步骤 5

选择"基本存储设备",如图 1.5 所示。

图 1.5 选择"基本存储设备"

步骤 6

选择基本存储设备后会提示存储设备数据会被删除,一定要注意备份好数据。选择"是,忽略所有数据(Y)",如图 1.6 所示。

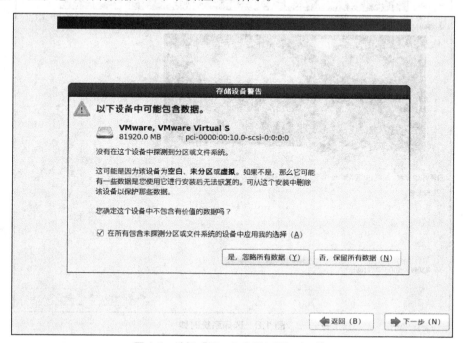

图 1.6 选择"是,忽略所有数据(Y)"

步骤 7

主机名默认就可以,以后可以修改,如图 1.7 所示。

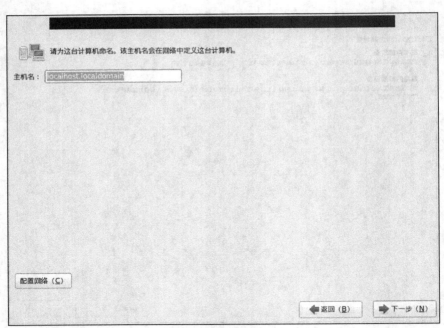

图 1.7 主机名默认

步骤 8

选择城市：亚洲/上海，系统时钟使用 UTC 时间(S)，如图 1.8 所示。

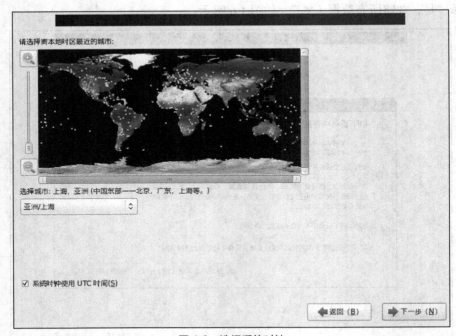

图 1.8 选择系统时钟

步骤 9

输入密码。密码尽量要符合密码复杂性要求，如果密码太简单，系统会提示"你的密码不够安全：过于简单化/系统化"，如图 1.9 所示。

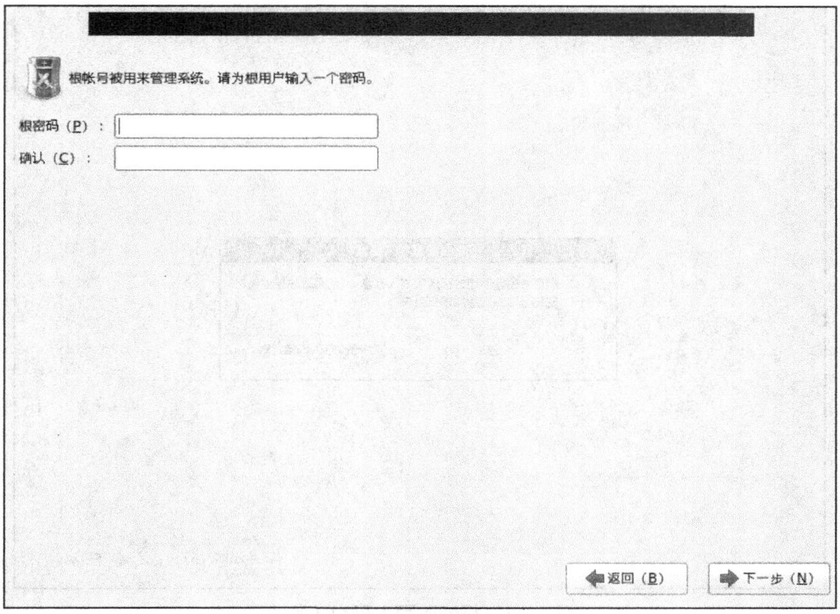

图 1.9　创建用户密码

步骤 10

可以选择"使用所有空间",也可以选择"创建自定义布局"或其他,选择"创建自定义布局"对初学者来说比较难,如图 1.10 所示。

图 1.10　选择安装类型

步骤 11

分区布局完成后,选择"将修改写入磁盘(W)",如图 1.11 所示。

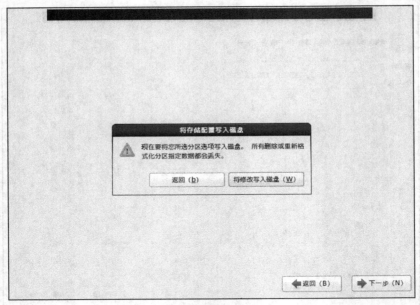

图 1.11　将修改写入磁盘（W）

步骤 12

根据需要选择系统安装。如果选择"Minimal"最小安装，安装完后系统没有加载图像界面，对于初学者来说操作不方便；选择"Desktop"桌面系统，比较适合初学者。本系统选择"Minimal Desktop"，如图 1.12 所示。

图 1.12　本系统选择"Minimal Desktop"

步骤 13

选择"Minimal Desktop"，按"下一步"按钮后，开始安装系统，此过程需要一些时间。安装后提示安装成功，如图 1.13 所示。

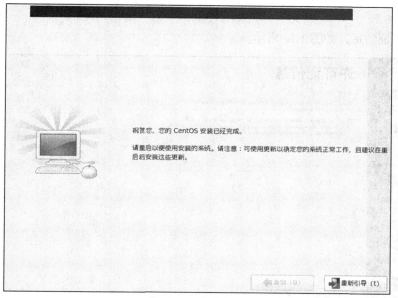

图 1.13 安装完成

步骤 14

安装完毕后重启系统，这时按任何键，可以对系统引导进行修改，比如可以设置以单用户模式启动系统，一般不需要按键盘，3 秒后自动进入系统，如图 1.14 所示。

图 1.14 安装完毕后重启系统

步骤 15

进入欢迎界面，如图 1.15 所示。

图 1.15 进入欢迎界面

步骤 16

显示许可证信息，如图 1.16 所示。

图 1.16　显示许可证信息

步骤 17

系统第一次启动后，会提示创建用户，可根据需要创建，这里直接前进不创建用户，如图 1.17 所示。

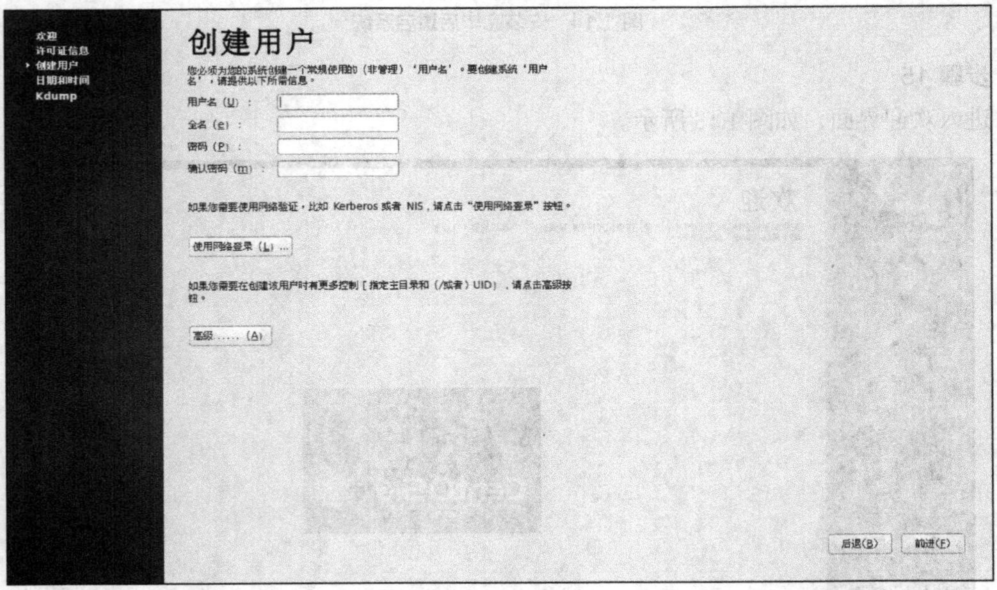

图 1.17　创建用户

步骤 18

设置日期和时间，如图 1.18 所示。

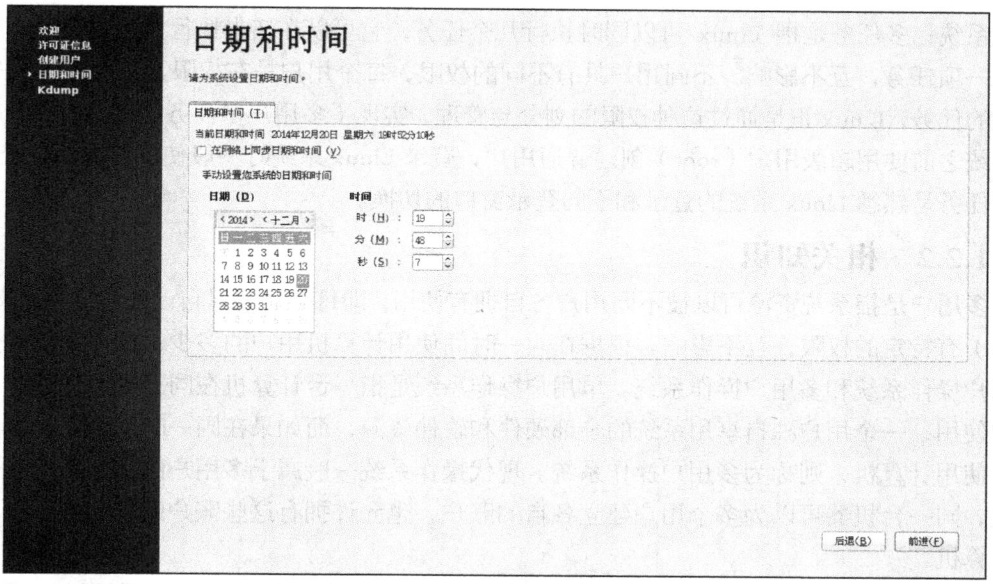

图 1.18 设置日期和时间

步骤 19

Kdump 是一个新的可信赖的内核崩溃转储机制。在启动时，Kdump 保留了一定数量的内存。系统内存较少时一般不要启用 Kdump。最后按"完成"，系统安装完毕，如图 1.19 所示。

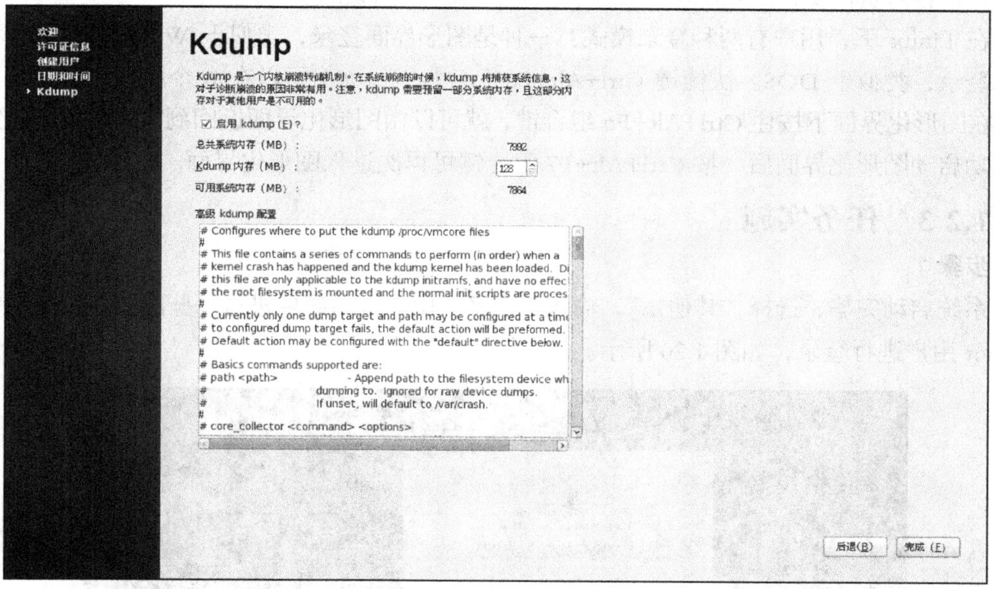

图 1.19 系统安装完毕

1.2 任务二 用户登录

1.2.1 任务描述

Linux 是一个多用户多任务的操作系统。多用户是指多个用户可以在同一时间使用一个计算机系统；多任务是指 Linux 可以同时执行几个任务，它可以在还未执行完一个任务时又执行另一项任务，互不影响。不同用户具有不同的权限，每个用户是在权限允许的范围内完成不同的任务，Linux 正是通过这种权限的划分与管理，实现了多用户多任务的运行机制。在使用系统之前要用超级用户（root）创建普通用户，登录 Linux 系统时一般使用普通用户登录。本节任务是熟悉 Linux 系统的登录和不同登录窗口的切换。

1.2.2 相关知识

多用户是指系统资源可以被不同用户各自拥有使用，即每个用户对自己的资源（如文件、设备）有特定的权限，互不影响。根据在同一时间使用计算机用户的多少，操作系统可分为单用户操作系统和多用户操作系统。单用户操作系统是指一台计算机在同一时间只能由一个用户使用，一个用户独自享用系统的全部硬件和软件资源，而如果在同一时间允许多个用户同时使用计算机，则称为多用户操作系统。现代操作系统一般属于多用户的操作系统，也就是说，同一台机器可以为多个用户建立各自的账户，也允许拥有这些账户的用户同时登录这台计算机。

Linux 实现多用户特性的关键在于，将所有系统调用在将数据准备好后通过一个接口（system_call）进入核心态，由核心态进行权限检查控制，并且保证资源的独占访问。在表面上看，系统调用就和其他的函数调用一样，只要结果符合预计的情况，应用程序就不能确定是否真正使用了内核，从而达到核心态切换对用户层透明的目的。这样的过程也就保证了每个用户进程对资源操作的互不影响，从而实现了 Linux 系统的多用户特性。

在 Linux 下，用户有两种登录模式：一种是图形界面登录，类似于 Windows；另一种是文本登录，类似于 DOS。快捷键 Ctrl+Alt+Fn（n=1、…、6）可以在 6 个虚拟终端中相互切换。在图形化界面下按住 Ctrl+Alt+Fn 组合键，就可以由图形化界面返回到文本界面。当第一次成功启动图形化界面后，按 Ctrl+Alt+F7 组合键可再次进入图形化界面。

1.2.3 任务实施

步骤 1

系统启动完毕，选择"其他..."，输入用户名。由于本系统要进行一些配置，使用超级用户 root 用户进行登录，如图 1.20 所示。

图 1.20 登录系统

步骤 2

用户登录窗口，如图 1.21 所示。

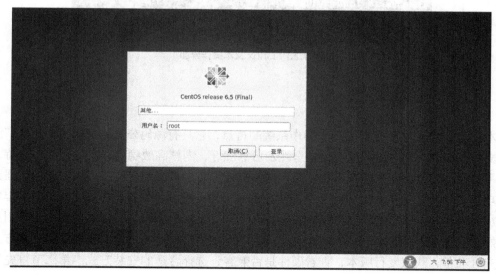

图 1.21 用户登录窗口

步骤 3

进入 Linux 系统操作界面，如图 1.22 所示。

图 1.22 进入 Linux 系统操作界面

步骤 4

按 Ctrl+Alt+Fn 组合键切换登录窗口，一般按 Ctrl+Alt+F1 组合键或 Ctrl+Alt+F7 组合键登录图形窗口，其他为文本窗口，如图 1.23 所示。

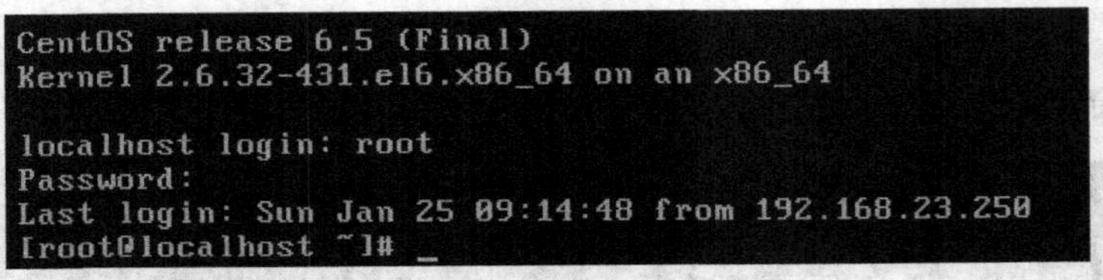

```
CentOS release 6.5 (Final)
Kernel 2.6.32-431.el6.x86_64 on an x86_64

localhost login: root
Password:
Last login: Sun Jan 25 09:14:48 from 192.168.23.250
[root@localhost ~]#
```

图 1.23　Linux 文本操作界面

1.3　本章小结

　　本章第一节首先介绍了 Linux 操作系统的特点，它具有源码公开，是自由软件；广泛的硬件支持，极强的平台可伸缩性；安全性及可靠性好，内核高效稳定；广泛的硬件支持，极强的平台可伸缩性；安全性及可靠性好，内核高效稳定；真正的多任务多用户；具有强大的网络功能；完全符合 POSIX 标准；具有丰富的图形用户界面等特点。CentOS 在 2014 年年初宣布加入 Red Hat 作为其开源和标准团队的一部分，与 Fedora 和 RHEL 共同协作，为用户提供更好的开源平台。安装 CentOS 前，必须要对硬件的基本要求、硬件的兼容性和安装方式等进行充分准备，获取发行版本，查看硬件是否兼容，选择合适的安装方式。第二节首先说明了 Linux 是一个多用户多任务的操作系统并介绍了多用户操作系统的特点。通过本章的学习重点要掌握 CentOS 的安装步骤和用户登录的方法。

第 2 章 CentOS 网络管理

CentOS 安装完成后，需要配置网络，实现网络互联。如何进行 CentOS 网络配置和管理成为我们迫切需要解决的问题。掌握 CentOS 基本的网络管理也是云计算集群维护和管理的基础。CentOS 网络的配置方式主要有三种：使用 NetworkManager 服务配置网络、使用 setup 命令配置网络、直接编辑网络配置文件。这三种网络配置的方式将在本章逐一介绍到。

CentOS 之所以具有较好的安全性及可靠性，与其自带的防火墙和安全策略密切相关。在众多的网络防火墙产品中，CentOS 上的防火墙软件特点显著，功能强大，而且源代码公开。这些优势是其他防火墙产品不可比拟的。虽然防火墙等安全设置是为了用户网络安全而存在，但是同时也带来网络功能的限制，增加了用户的操作和管理难度，本章也将介绍如何关闭防火墙等安全设置。

2.1 任务一 网络配置

2.1.1 任务描述

CentOS 具有强大的网络功能，提供方便用户操作的网络配置工具，我们要正确使用工具配置网络，也可以直接编辑网络配置文件配置网络。本节任务主要完成 CentOS 网络的配置。

2.1.2 相关知识

1．NetworkManager

NetworkManager 由一个管理系统网络连接，将其状态通过 D-BUS 进行报告的后台服务，以及一个允许用户管理网络连接的客户端程序组成。开发 NetworkManager 的初衷是简化网络连接的工作，让桌面本身和其他应用程序能感知网络。NetworkManager 让许多用户在使用主流、标准的网络配置时变得简便，但未必适应那些不常见的配置。

2．子网掩码

子网掩码（subnet mask）又叫网络掩码、地址掩码，它是一种用来指明一个 IP 地址的哪些位标识的是主机所在的子网，以及哪些位标识的是主机的位掩码。子网掩码只有一个作用，就是将某个 IP 地址划分成网络地址和主机地址两部分。

3．网关

网关（Gateway）又称网间连接器、协议转换器。默认网关在网络层上以实现网络互联，是最复杂的网络互联设备，仅用于两个高层协议不同的网络互联。网关既可以用于广域网互联，也可以用于局域网互联。网关是一种充当转换重任的计算机系统或设备。

2.1.3 任务实施

方法一： 使用 NetworkManager 服务配置网络。

步骤 1

打开编辑连接，如图 2.1 所示。

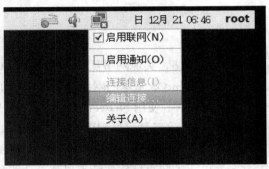

图 2.1 打开编辑连接

步骤 2

进入网络连接编辑窗口，如图 2.2 所示。

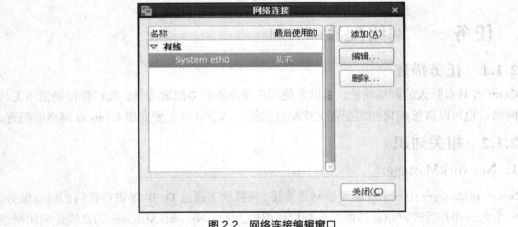

图 2.2 网络连接编辑窗口

步骤 3

IPv4 设置，如图 2.3 所示。

步骤 4

由于 hadoop 不支持 NetworkManger，建议关闭 NetworkManager 服务。停止 NetworkManager 服务，运行级别 2、3、4、5 关闭，下次开机服务不会启动。操作命令如下。

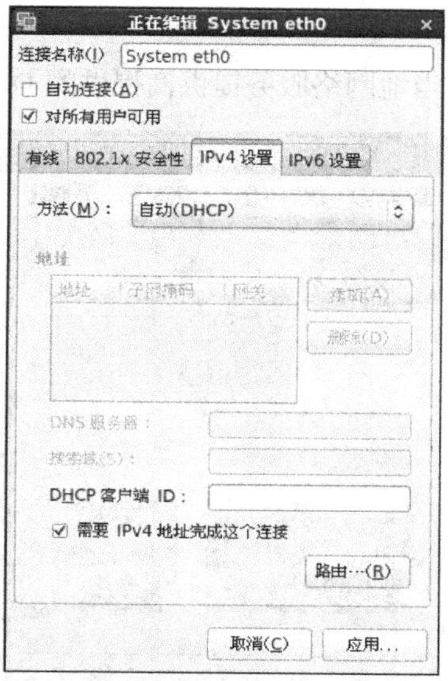

图 2.3 IPv4 设置

```
[root@localhost ~]# service NetworkManager stop
停止 NetworkManager 守护进程：                            [确定]
[root@localhost ~]# chkconfig --level 23456 NetworkManager off
[root@localhost ~]# chkconfig --list NetworkManager
NetworkManager   0:关闭   1:关闭   2:关闭   3:关闭   4:关闭   5:关闭   6:关闭
[root@localhost ~]#
```

方法二：使用 setup 命令配置网络。

步骤 1

运行 setup 命令，打开 setup 窗口，如图 2.4 所示。

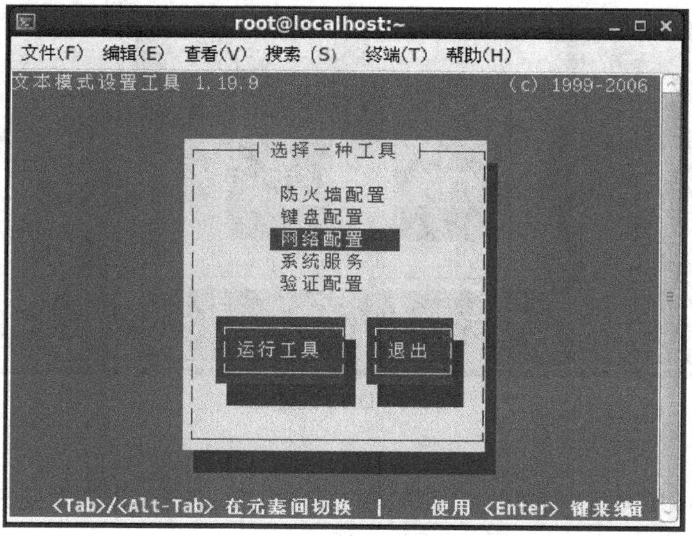

图 2.4 打开 setup 窗口

步骤 2

配置 DNS。DNS 一般使用当地网络服务提供商提供的 DNS 的 IP 地址,如图 2.5 所示。

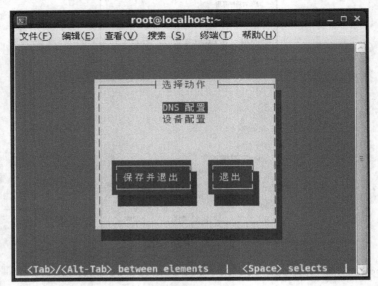

图 2.5 配置 DNS

步骤 3

DNS 配置窗口,如图 2.6 所示。

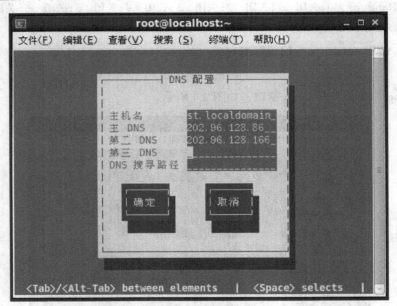

图 2.6 DNS 配置窗口

步骤 4

配置以太网的 IP 地址。这里要注意以太网的名称和设备号,以太网设备号从 eth0 开始编号,以太网名称和设备号建议相同,如图 2.7 所示。

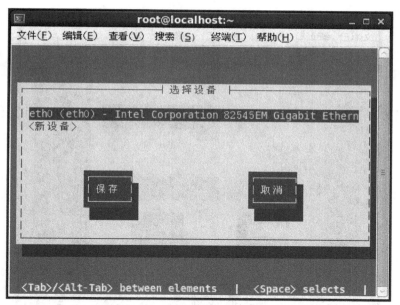

图 2.7 以太网 IP 配置

有时由于网卡的更换，导致以太网设备号和网卡 MAC 地址的变化，可使用 ifconfig –a 查看以太网设备号及 MAC 地址，如下所示。

```
[root@localhost ~]# ifconfig -a
eth1      Link encap:Ethernet  HWaddr 00:0C:29:09:D5:6E
          inet6 addr: fe80::20c:29ff:fe09:d56e/64 Scope:Link
          UP BROADCAST RUNNING MULTICAST  MTU:1500  Metric:1
          RX packets:3752589 errors:0 dropped:0 overruns:0 frame:0
          TX packets:96624 errors:0 dropped:0 overruns:0 carrier:0
          collisions:0 txqueuelen:1000
          RX bytes:1575680079 (1.4 GiB)  TX bytes:36772302 (35.0 MiB)

lo        Link encap:Local Loopback
          inet addr:127.0.0.1  Mask:255.0.0.0
          inet6 addr: ::1/128 Scope:Host
          UP LOOPBACK RUNNING  MTU:16436  Metric:1
          RX packets:141814 errors:0 dropped:0 overruns:0 frame:0
          TX packets:141814 errors:0 dropped:0 overruns:0 carrier:0
          collisions:0 txqueuelen:0
          RX bytes:182673081 (174.2 MiB)  TX bytes:182673081 (174.2 MiB)
```

步骤 5

默认以太网自动获取 IP，使用 DHCP 有 "*" 号，表示网络使用 DHCP 分配的 IP 地址，使用空格键进行选择切换，如图 2.8 所示。

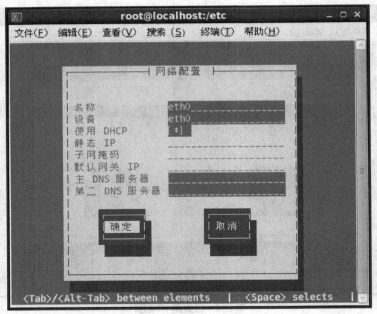

图 2.8 默认以太网自动获取 IP

步骤 6

设置固定 IP，这里的 DNS 服务器可以不设置，如图 2.9 所示。

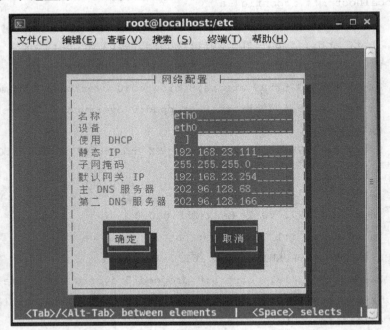

图 2.9 设置固定 IP

步骤 7

保存并退出，如图 2.10 和图 2.11 所示。

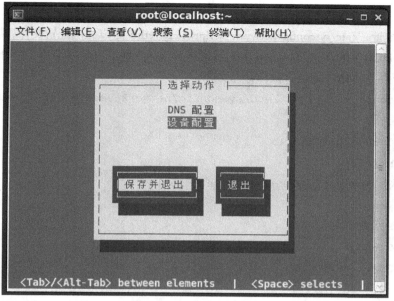

图 2.10 保存并退出

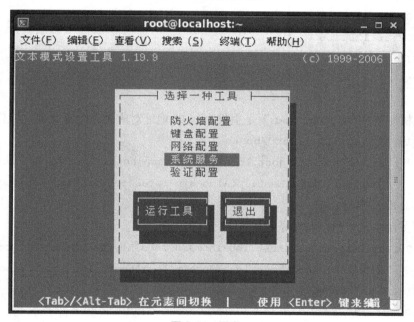

图 2.11 退出

方法三：直接编辑网络配置文件。

步骤 1

配置 DNS，"/etc/resolv.conf"是 DNS 配置文件，最多设置三台 DNS 服务器，配置如下。

```
[root@localhost ~]# vi /etc/resolv.conf
nameserver 202.96.128.68
nameserver 202.96.128.166
```

步骤 2

配置 IP 地址,"/etc/sysconfig/network-scripts/ifcfg-eth0"是以太网配置文件,其中 DEVICE=eht0 和 HWADDR =00:0c:29:f4:be:4b 中的设备号和 MAC 地址一定要准确,使用 Ifconfig –a 查看以太网设备号和 MAC 地址。操作命令如下。

```
[root@localhost ~]# vi /etc/sysconfig/network-scripts/ifcfg-eth0
DEVICE=eth0
HWADDR=00:0c:29:f4:be:4b
TYPE=Ethernet
UUID=c9383bcd-8dff-4d4b-99a2-250b899f5513
ONBOOT=yes
NM_CONTROLLED=no
BOOTPROTO=none
IPADDR=192.168.23.111
NETMASK=255.255.255.0
GATEWAY=192.168.23.254
DNS1=202.96.128.68
DNS2=202.96.128.166
IPV6INIT=no
USERCTL=no
```

步骤 3

设置主机配置文件,"/etc/hosts"是主机名与 IP 地址对应关系配置文件。操作命令如下。

```
[root@localhost ~]#vi /etc/hosts
127.0.0.1 localhost localhost.localdomain localhost4 localhost4.localdomain4
::1       localhost localhost.localdomain localhost6 localhost6.localdomain6
192.168.23.111 node1
```

如果有多台主机,可以写多条记录。

```
127.0.0.1 localhost localhost.localdomain localhost4 localhost4.localdomain4
::1       localhost localhost.localdomain localhost6 localhost6.localdomain6
192.168.23.111 node1
192.168.23.112 node2
192.168.23.113 node3
192.168.23.114 node4
```

步骤 4

临时修改主机名,操作命令如下。

```
[root@localhost ~]# hostname node1
```

步骤 5

永久性修改主机名,"/etc/sysconfig/network"是主机名配置文件。操作命令如下。

```
[root@localhost ~]# vi /etc/sysconfig/network
NETWORKING=yes
HOSTNAME=localhost.localdomain
```

步骤 6

网络配置后，重启 network 服务，或者先停止服务后再启动服务。操作命令如下。

```
root@localhost ~]# service network restart
```

显示如下，表示设置成功。

```
Shutting down interface eth0:                              [  OK  ]
Shutting down loopback interface:                          [  OK  ]
Bringing up loopback interface:                            [  OK  ]
Bringing up interface eth0: Determining if ip address 192.168.23.111 is already
 in use for device eth0...
 [  OK  ]
```

步骤 7

测试网络是否连通。操作命令如下。显示表示配置正确。使用 ping 测试，需按 Ctrl+C 组合键中断测试。

```
[root@localhost ~]# ping www.baidu.com
PING www.a.shifen.com (180.97.33.108) 56(84) bytes of data.
64 bytes from 180.97.33.108: icmp_seq=1 ttl=53 time=27.8 ms
64 bytes from 180.97.33.108: icmp_seq=2 ttl=53 time=27.5 ms
^C
--- www.a.shifen.com ping statistics ---
2 packets transmitted, 2 received, 0% packet loss, time 1833ms
rtt min/avg/max/mdev = 27.540/27.682/27.824/0.142 ms
```

2.2 任务二 更改以太网卡名称

2.2.1 任务描述

Linux 网络配置中一般用 eth0、eth1、eth2 标准以太网卡的名称。当使用 Linux 更换了以太网卡，或者将虚拟机从一台计算机复制到另一台计算机时，由于设备以太网卡 mac 地址改变，但是系统配置文件/etc/udev/rules.d/70-persistent-net.rules 中仍然保留了旧以太网卡的内容，新以太网卡则被识别为 eth1。为了操作习惯，我们把本机的以太网卡名称重新改为 eth0。本节任务是了解网络设备和网络 MAC 地址，掌握以太网卡名称的更改方法，完成以太网卡名称的修改。

2.2.2 相关知识

1. 网络设备

Linux 的以太网卡通常命名为 eth0、eth1 等。当一个以太网卡被侦测到时，它会被指定为第一个可用的接口卡名字，通常为 eth0。但 Linux 的网络设备并不一定会与/dev 目录下的特殊文件联系在一起，所以如果在/dev 内找不到它们时请不要觉得惊讶。

2．MAC 地址

MAC（Media Access Control 或者 Medium Access Control）地址，意译为媒体访问控制，或称为物理地址、硬件地址，用来定义网络设备的位置。在 OSI 模型中，第三层网络层负责 IP 地址，第二层数据链路层则负责 MAC 地址。因此一个主机会有一个 MAC 地址，而每个网络位置会有一个专属于它的 IP 地址。

2.2.3 任务实施

步骤 1

使用 ifconfig –a 查看 eth1 的 MAC，操作命令如下。

```
[root@localhost ~]# ifconfig -a
eth1    Link encap:Ethernet  HWaddr 00:0C:29:09:D5:6E
        inet6 addr: fe80::20c:29ff:fe09:d56e/64 Scope:Link
        UP BROADCAST RUNNING MULTICAST  MTU:1500  Metric:1
        RX packets:3752589 errors:0 dropped:0 overruns:0 frame:0
        TX packets:96624 errors:0 dropped:0 overruns:0 carrier:0
        collisions:0 txqueuelen:1000
        RX bytes:1575680079 (1.4 GiB)  TX bytes:36772302 (35.0 MiB)
```

其中 eth1 的 MAC 是 00:0C:29:09:D5:6E。

步骤 2

修改"/etc/udev/rules.d/70-persistent-net.rules"，删除含有"Name="eth0""的一行，把含有"Name="eth1""一行中"Name="eth1""改为"Name="eth0""，该行显示的 MAC 地址是 00:0C:29:09:D5:6E，与上面相同。操作命令如下。

```
[root@localhost ~]# vi /etc/udev/rules.d/70-persistent-net.rules
```

把下面这段的内容：

```
# PCI device 0x8086:0x100f (e1000)
SUBSYSTEM=="net",                       ACTION=="add",           DRIVERS=="?*",
ATTR{address}=="00:0c:29: f4:be:4b", ATTR{type}=="1", KERNEL=="eth*", NAME="eth0"

# PCI device 0x8086:0x100f (e1000)
SUBSYSTEM=="net",                       ACTION=="add",           DRIVERS=="?*",
 ATTR{address}=="00:0c:29: 09:d5:6e", ATTR{type}=="1", KERNEL=="eth*", NAME="eth1"
```

改为如下所示内容：

```
# PCI device 0x8086:0x100f (e1000)
SUBSYSTEM=="net",                       ACTION=="add",           DRIVERS=="?*",
ATTR{address}=="00:0c:29: 09:d5:6e", ATTR{type}=="1", KERNEL=="eth*", NAME="eth0"
```

步骤 3

修改 MAC 地址，编辑"/etc/sysconfig/network-scripts/ifcfg-eth0"，修改 HWADDR 中的 MAC 地址。操作命令如下。

```
[root@localhost ~]# vi /etc/sysconfig/network-scripts/ifcfg-eth0
```

把原来的 HWADDR= 00:0c:29:f4:be:4b，改为 HWADDR=00:0C:29:09:d5:6e，修改后 ifcfg-eth0 配置内容如下。

```
DEVICE=eth0
HWADDR=00:0c:29:09:d5:6e
TYPE=Ethernet
UUID=c9383bcd-8dff-4d4b-99a2-250b899f5513
ONBOOT=yes
NM_CONTROLLED=no
BOOTPROTO=dhcp
IPV6INIT=no
USERCTL=no
```

BOOTPROTO=none 表示网络 IP 地址固定设置，BOOTPROTO dhcp 表示网络自动获取 IP 地址。

2.3 任务三 关闭安全设置

2.3.1 任务描述

Linux 安全级别相对较高，Linux 自带的 iptables 能很好保护系统安全，SELinux 能防止一些非法操作，但也增加了我们操作上的限制。本节任务是了解 SELinux 的作用，了解防火墙的作用，掌握防火墙的关闭，掌握 SELinux 安全设置。

2.3.2 相关知识

防火墙等安全设置是一项协助确保信息安全的设备，会依照特定的规则，允许或是限制传输的数据通过。这些安全设置的作用就是保护用户的网络免受非法用户的侵入，虽然防火墙等安全设置是为了用户网络安全而存在，但是同时也限制了一些网络功能。

SELinux(Security-Enhanced Linux) 是美国国家安全局（NSA）对于强制访问控制的实现，是 Linux 历史上最杰出的新安全子系统。NSA 在 Linux 社区的帮助下开发了一种访问控制体系，在这种访问控制体系的限制下，进程只能访问那些它的任务中所需要文件。

2.3.3 任务实施

步骤 1

关闭 SELinux，操作命令如下。

```
[root@localhost ~]# setenforce 0
[root@localhost ~]# vi /etc/selinux/config
```

将"/etc/selinux/config"配置文件中"SELINUX=enforcing"改为"SELINUX=disabled"。

修改后的结果如下。

```
# This file controls the state of SELinux on the system.
# SELINUX= can take one of these three values:
#     enforcing - SELinux security policy is enforced.
#     permissive - SELinux prints warnings instead of enforcing.
#     disabled - No SELinux policy is loaded.
SELINUX=disabled
# SELINUXTYPE= can take one of these two values:
#     targeted - Targeted processes are protected,
#     mls - Multi Level Security protection.
SELINUXTYPE=targeted
```

步骤 2

关闭防火墙，操作命令如下。

```
[root@localhost ~]# service iptables stop
```

显示如下信息，表示防护墙已经关闭。

```
[root@localhost ~]# service iptables stop
iptables：将链设置为政策 ACCEPT：nat mangle filter      [确定]
iptables：清除防火墙规则：                              [确定]
iptables：正在卸载模块：                                [确定]
```

步骤 3

设置防火墙开机不自动启动，操作命令如下。

```
[root@localhost ~]# chkconfig --level 2345 iptables off
[root@localhost ~]# chkconfig --list iptables
iptables        0:关闭  1:关闭  2:关闭  3:关闭  4:关闭  5:关闭  6:关闭
```

2.4 本章小结

本章第一节介绍了网络配置的相关术语、网络配置文件和网络的基本命令，接着在任务实施部分介绍了 CentOS 网络配置常用的三种方式：使用 NetworkManager 服务配置网络、使用 setup 命令配置网络和直接编辑网络配置文件。第二节介绍了更改以太网设备号的相关知识和基本方法。第三节介绍了 Linux 防火墙的基本作用，虽然防火墙加强了安全，但是防火墙往往限制了用户的基本操作。通过本节的学习理解 SELinux 的作用，了解防火墙的作用，掌握防火墙的关闭方法，掌握 SELinux 安全设置方法。

第 3 章 虚拟化技术

虚拟化是云计算中主要支撑技术之一。虚拟化将应用程序和数据在不同层次以不同的面貌展现，这样有助于使用者、开发及维护人员方便地使用、开发及维护这些应用程序及数据。虚拟化允许 IT 部门添加、减少移动硬件和软件到他们想要的地方。虚拟化为组织带来灵活性，从而改善 IT 运维和减少成本支出。云计算和虚拟化并非捆绑技术，二者同时使用仍可正常运行并实现优势互补。云计算和虚拟化二者交互工作，云计算解决方案依靠并利用虚拟化提供服务，而那些尚未部署云计算解决方案的公司仍然可以利用端到端虚拟化从内部基础设施中获得更佳的投资回报和收益。

云计算将各种 IT 资源以服务的方式通过互联网交付给用户，然而虚拟化本身并不能给用户提供自服务层。没有自服务层，就不能提供计算服务。云计算模型允许终端用户自行提供自己的服务器、应用程序和包括虚拟化等其他的资源，这反过来又能使企业最大程度地处理自身的计算资源，并且需要系统管理员为终端用户提供虚拟机。

3.1 任务一 配置虚拟化网络

3.1.1 任务描述

KVM 虚拟机网络连接方式主要有两种：NAT 方式和 Bridge 方式。虚拟机安装完成后，需要为其设置网络接口，以便和主机网络、客户机之间进行网络通信。本节任务完成 Bridge 方式的配置。通过本节的学习，读者将了解虚拟网络的概念，了解 NAT 方式和 Bridge 方式的工作方式，掌握 Bridge 方式的配置。

3.1.2 相关知识

KVM 客户机网络连接有两种方式。

- 用户网络（User Networking）：让虚拟机访问主机、互联网或本地网络上的资源的简单方法，但是不能从网络或其他的客户机访问客户机，该配置方式为 NAT 方式。
- 虚拟网桥（Virtual Bridge）：这种方式要比用户网络复杂一些，但是设置好后客户机与互联网、客户机与主机之间的通信都很容易，该配置方式为 Bridge 方式。

1. NAT 方式

NAT 方式是 KVM 安装后的默认方式，支持桌面主机虚拟化。它支持主机与虚拟机的互访，同时也支持虚拟机访问互联网，但不支持外界访问虚拟机。NAT 方式如图 3.1 所示，从图上可以看出，虚拟接口和物理接口之间没有直接连接关系，虚拟网络与物理网络通过建立映射，虚拟机才能访问外部世界，外部无法从网络上定位和访问虚拟主机。

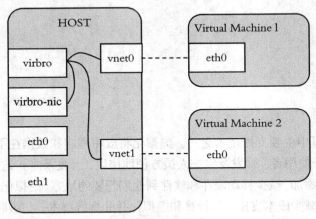

图 3.1 NAT 方式

2. Bridge 方式

Bridge 方式即虚拟网桥的网络连接方式，使客户机和子网里面的机器能够互相通信。它可以使虚拟机成为网络中具有独立 IP 的主机。桥接网络（也叫物理设备共享）被用于将一个物理设备复制到一台虚拟机。网桥多用作高级设置，特别是主机多个网络接口的情况。Bridge 网桥的基本原理就是创建一个桥接接口 br0，在物理网卡和虚拟网络接口之间传递数据。方式如图 3.2 所示。

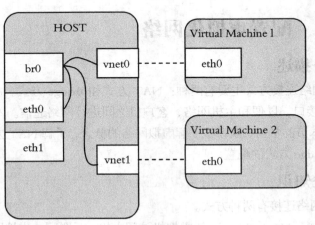

图 3.2 Bridge 方式

3.1.3 任务实施

步骤 1

关闭 selinux，关闭防火墙。

步骤 2

进入网络配置目录/etc/sysconfig/network-scripts/，复制配置文件。操作命令如下。

```
[root@localhost ~]# cd /etc/sysconfig/network-scripts/
[root@localhost network-scripts]# cp ifcfg-eth0 ifcfg-br0
```

编辑以太网卡配置文件 ifcfg-eth0，添加最后一行 BRIDGE=br0。操作命令如下。

```
[root@localhost network-scripts]# vi ifcfg-eth0
DEVICE=eth0
HWADDR=00:0c:29:09:d5:6e
TYPE=Ethernet
UUID=c9383bcd-8dff-4d4b-99a2-250b899f5513
ONBOOT=yes
NM_CONTROLLED=no
BOOTPROTO=none
IPADDR=192.168.23.100
NETMASK=255.255.255.0
GATEWAY=192.168.23.254
IPV6INIT=no
USERCTL=no
BRIDGE=br0
```

修改桥接配置文件 ifcfg-br0，操作命令如下。

```
[root@localhost network-scripts]# vi ifcfg-br0
DEVICE=br0
TYPE=bridge
ONBOOT=yes
NM_CONTROLLED=no
BOOTPROTO=none
IPADDR=192.168.23.100
NETMASK=255.255.255.0
GATEWAY=192.168.23.254
IPV6INIT=no
USERCTL=no
```

步骤 3

重启网络服务，操作命令如下。

```
[root@localhost network-scripts]# service network restart
```

重启后显示如下信息，表示 br0 没有启动。

正在关闭接口 eth0：	[确定]
关闭环回接口：	[确定]
弹出环回接口：	[确定]

弹出界面 br0：设备 br0 似乎不存在，初始化操作将被延迟

[失败]

弹出界面 eth0： [确定]

启动 br0 设备，操作命令及运行结果如下。

```
[root@localhost network-scripts]# ifup br0
Determining if ip address 192.168.23.100 is already in use for device br0...
```

说明 br0 的 IP 替代 eht0 的 IP 配置。

步骤 4

设置开机自动启动 br0 设备，操作命令如下。

```
[root@localhost network-scripts]# vi /etc/rc.d/rc.local
```

在文件末尾添加如下内容。

```
ifup br0
```

步骤 5

查看网络生效信息，操作命令如下。

```
[root@localhost network-scripts]# ifconfig -a
```

显示如下内容，表示 br0 的 IP 替代 eht0 的 IP 配置。

```
br0       Link encap:Ethernet  HWaddr 00:0C:29:09:D5:6E
          inet addr:192.168.23.100  Bcast:192.168.23.255  Mask:255.255.255.0
          inet6 addr: fe80::20c:29ff:fe09:d56e/64 Scope:Link
          UP BROADCAST RUNNING MULTICAST  MTU:1500  Metric:1
          RX packets:896834 errors:0 dropped:0 overruns:0 frame:0
          TX packets:170564 errors:0 dropped:0 overruns:0 carrier:0
          collisions:0 txqueuelen:0
          RX bytes:786138293 (749.7 MiB)  TX bytes:76227924 (72.6 MiB)


eth0      Link encap:Ethernet  HWaddr 00:0C:29:09:D5:6E
          inet6 addr: fe80::20c:29ff:fe09:d56e/64 Scope:Link
          UP BROADCAST RUNNING MULTICAST  MTU:1500  Metric:1
          RX packets:12384052 errors:0 dropped:0 overruns:0 frame:0
          TX packets:200652 errors:0 dropped:0 overruns:0 carrier:0
          collisions:0 txqueuelen:1000
          RX bytes:4372315301 (4.0 GiB)  TX bytes:78260836 (74.6 MiB)
```

步骤 6

查看桥接情况，操作命令及运行结果如下。

```
[root@localhost network-scripts]# brctl show
bridge name     bridge id               STP enabled     interfaces
br0             8000.000c2909d56e       no              eth0
```

表示桥接成功。

3.2 任务二 安装虚拟化软件包

3.2.1 任务描述

KVM（Kernel-based Virtual Machine）是一种针对 Linux 内核的虚拟化基础架构，它支持具有硬件虚拟化扩展的处理器上的原生虚拟化。本节任务是完成 KVM 虚拟化软件的安装，通过本节的学习，用户将进一步了解虚拟化概念，了解虚拟化技术。

3.2.2 相关知识

1. 什么是虚拟化技术

虚拟化是指计算机元件在虚拟的基础上而不是真实的基础上运行。虚拟化技术可以扩大硬件的容量，简化软件的重新配置过程。CPU 的虚拟化技术可以以单 CPU 模拟多 CPU 并行，允许一个平台同时运行多个操作系统，并且应用程序都可以在相互独立的空间内运行而互不影响，从而显著提高计算机的工作效率。虚拟化是一个广义的术语，是一个简化管理、优化资源的解决方案，如同现在空旷、通透的写字楼，整个楼层几乎看不到墙壁，用户可以用同样的成本构建出更加自主适用的办公空间，进而节省成本，发挥空间最大利用率。这种把有限的固定的资源根据不同需求进行重新规划以达到最大利用率的思路，在 IT 领域就叫作虚拟化技术。

（1）硬件虚拟化——虚拟化技术的革命。

CPU 的虚拟化技术是一种硬件方案，支持虚拟技术的 CPU 带有经特别优化过的指令集来控制虚拟过程，通过这些指令集，虚拟机可以很容易提高性能，相比纯软件的虚拟化技术有很大程度上的提高。

（2）纯软件的虚拟化技术。

在纯软件虚拟化解决方案中，虚拟机中的操作系统其实是真实操作系统下的一个应用程序，因此，虚拟操作系统上的应用程序到实际操作系统就要比通常应用程序多经过一个通信层。

虚拟化技术分为：平台虚拟化（Platform Virtualization），针对计算机和操作系统的虚拟化；资源虚拟化（Resource Virtualization），针对特定的系统资源的虚拟化，比如内存、存储、网络资源等；应用程序虚拟化（Application Virtualization），包括仿真、模拟、解释技术等。

我们通常所说的虚拟化主要是指平台虚拟化技术，通过使用控制程序（Control Program，也被称为 Virtual Machine Monitor 或 Hypervisor），隐藏特定计算平台的实际物理特性，为用户提供抽象的、统一的、模拟的计算环境（称为虚拟机）。虚拟机中运行的操作系统被称为客户机操作系统（Guest OS），运行虚拟机监控器的操作系统被称为主机操作系统（Host OS），当然某些虚拟机监控器可以脱离操作系统直接运行在硬件之上（如 VMWARE 的 ESX 产品）。运行虚拟机的真实系统我们称之为主机系统。

可以想象一下，未来的虚拟化发展将会是多元化的，包括服务器、存储、网络等更多的元素，用户将无法分辨哪些是虚，哪些是实。虚拟化将改变现在的传统 IT 架构，而且将互联网中的所有资源全部连在一起，形成一个大的计算中心，而用户却不用关心所有这一切，而只需关心提供给自己的服务是否正常。虽然虚拟化技术前看好，但是，这一过程还有很长的路要走,因为还没有哪种技术是不存在潜在缺陷甚至陷阱的。可以相信,虚拟化技术将会成为未来的主要发展方向。

2. KVM 虚拟化软件

KVM(Kernel-based Virtual Machine)是一种针对 Linux 内核的虚拟化基础架构,它支持具有硬件虚拟化扩展的处理器上的原生虚拟化。最初,它支持 x86 处理器,但现在广泛支持各种处理器和操作系统,包括 Linux、BSD、Solaris、Windows、Haiku、ReactOS 和 AR-OS 等。基于内核的虚拟机(KVM)是针对包含虚拟化扩展(Intel VT 或 AMD-V)的 x86 硬件上的 Linux 的完全原生的虚拟化解决方案。对半虚拟化(Paravirtualization)的有限支持也可以通过半虚拟网络驱动程序的形式用于 Linux 和 Windows Guest 系统。

尽管 KVM 是一个相对较新的虚拟机管理程序,但这个随主流 Linux 内核发布的轻量型模块提供简单的实现和对 Linux 重要任务的持续支持。KVM 使用很灵活,Guest 操作系统与集成到 Linux 内核中的虚拟机管理程序通信,以直接寻址硬件,无需修改虚拟化的操作系统。这使得 KVM 成为更快的虚拟机解决方案。KVM 的补丁与 Linux 内核兼容,KVM 在 Linux 内核本身内实现,进而简化对虚拟化进程的控制,但是没有成熟的工具可用于 KVM 服务器的管理,KVM 仍然需要改进虚拟网络的支持、虚拟存储的支持,并且增强安全性、高可用性、容错、电源管理、HPC/实时支持、虚拟 CPU 可伸缩性、跨供应商兼容性、VM 可移植性。

3.2.3 任务实施

方法一:使用添加/删除软件安装软件包。

步骤 1

运行添加删除软件,打开"系统—管理—添加/删除软件",如图 3.3 所示。

图 3.3 软件管理工具

选择"添加/删除软件"后,选择"虚拟化、虚拟化客户端、虚拟化平台"安装,也可以完整安装所有虚拟化软件包。

步骤 2

安装虚拟化,如图 3.4 所示。

步骤 3

虚拟化附加软件包必须安装,如图 3.5 所示。

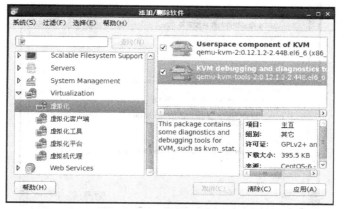

图 3.4 软件管理窗口

图 3.5 安装虚拟化附加软件包

步骤 4

安装虚拟化客户端，如图 3.6 所示。

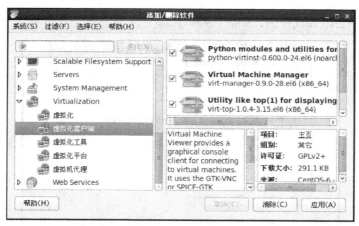

图 3.6 安装虚拟化客户端

步骤 5

安装虚拟化客户端附加软件包，如图 3.7 所示。

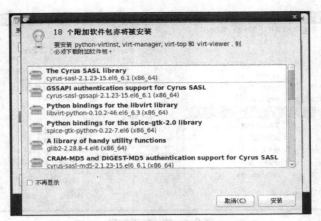

图 3.7 安装虚拟化客户端附加软件包

步骤 6

安装虚拟化平台，如图 3.8 所示。

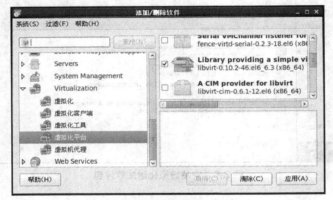

图 3.8 安装虚拟化平台

步骤 7

安装虚拟化平台附加软件包，如图 3.9 所示。

图 3.9 安装虚拟化平台附加软件包

方法二：使用命令安装软件包。

如果没有图形界面，可以使用 rpm 命令安装虚拟化软件包，具体批处理文件在资料库里。

步骤 1

安装虚拟化，操作命令如下。

```
#Userspace component of KVM （11个附加软件包）
rpm -ivh glusterfs-libs-3.4.0.36rhs-1.el6.x86_64.rpm
rpm -ivh glusterfs-3.4.0.36rhs-1.el6.x86_64.rpm
rpm -ivh vgabios-0.6b-3.7.el6.noarch.rpm
……
#KVM debugging and diagnostics tools （1个软件包）
rpm -ivh qemu-kvm-tools-0.12.1.2-2.415.el6.x86_64.rpm
```

步骤 2

安装虚拟化客户端，操作命令如下。

```
#Python modules and utilities for installing virtual machines （11个附加软件包）
rpm -ivh cyrus-sasl-md5-2.1.23-13.el6_3.1.x86_64.rpm
rpm -ivh yajl-1.0.7-3.el6.x86_64.rpm
rpm -ivh libvirt-client-0.10.2-29.el6.x86_64.rpm
……
#Virtual Machine Manager （21个附加软件包）
rpm -ivh yajl-1.0.7-3.el6.x86_64.rpm
rpm -ivh libcacard-0.15.0-2.el6.x86_64.rpm
rpm -ivh celt051-0.5.1.3-0.el6.x86_64.rpm
……
#Utility like top(1) for displaying virtualization stats （10个附加软件包）
rpm -ivh cyrus-sasl-md5-2.1.23-13.el6_3.1.x86_64.rpm
rpm -ivh yajl-1.0.7-3.el6.x86_64.rpm
rpm -ivh libvirt-client-0.10.2-29.el6.x86_64.rpm
……
#Virtual Machine viewer （17个附加软件包）
rpm -ivh yajl-1.0.7-3.el6.x86_64.rpm
rpm -ivh nc-1.84-22.el6.x86_64.rpm
rpm -ivh celt051-0.5.1.3-0.el6.x86_64.rpm
……
```

步骤 3

安装 libvirt 工具，操作命令如下。

```
rpm -ivh radvd-1.6-1.el6.x86_64.rpm
rpm -ivh ebtables-2.0.9-6.el6.x86_64.rpm
```

```
rpm -ivh cyrus-sasl-md5-2.1.23-13.el6_3.1.x86_64.rpm
......
```

步骤 4

关闭 NAT 网络,关闭 NAT 网络是为了减少 NAT 的影响。操作命令如下。

```
[root@localhost ~]# service libvirtd start
[root@localhost ~]# virsh net-destroy default
[root@localhost ~]# virsh net-undefine default
[root@localhost ~]# service libvirtd restart
```

步骤 5

安装客户端工具,是为了方便远程调用虚拟机。

(1) 安装 tigervnc 客户端,操作命令如下。

```
[root@localhost ~]# cd /media/CentOS_6.5_Final/Packages/
[root@localhost Packages]# rpm -ivh tigervnc-1.1.0-5.el6_4.1.x86_64.rpm
Preparing...              ########################################### [100%]
   1:tigervnc             ########################################### [100%]
```

(2) 安装 SPICE 客户端,操作命令如下。

```
[root@localhost Packages]# rpm -ivh spice-client-0.8.2-15.el6.x86_64.rpm
Preparing...              ########################################### [100%]
   1:spice-client         ########################################### [100%]
```

3.3 任务三 虚拟系统管理器的使用

3.3.1 任务描述

虚拟化系统管理器 virt-manager 使用方便直观,适合初学者。本节任务是使用虚拟化系统管理器 virt-manager 完成虚拟机的安装。通过本节的学习,读者将了解虚拟机安装方式,掌握虚拟机虚拟设备的设置,掌握虚拟机磁盘空间的分配,掌握虚拟化系统管理器 virt-manager 的使用。

3.3.2 相关知识

随着虚拟化的引入,物理主机得以摆脱单一实例操作系统的禁锢。我们通过多个操作系统用作虚拟机来有效地复用我们的主机。但是,一个主机上的操作系统越密集,就越会增加管理需求。这种管理问题的一个解决方案是 Virtual Machine Manager,或称为 virt-manager。

Virtual Machine Manager (virt-manager) 是一个轻量级应用程序套件,形式为一个管理虚拟机的命令行或图形用户界面(GUI)。除了提供对虚拟机的管理功能之外,virt-manager 还通过一个嵌入式虚拟网络计算(VNC)客户端查看器为 Guest 虚拟机提供一个完整图形控制台。作为一个应用程序套件,virt-manager 包括了一组常见的虚拟化管理工具。这些工具已在表 3.1 中列出,包括虚拟机构造、克隆、映像制作和查看。virsh 实用程序不是 virt-manager 包的一部分,但它本身就具有很重要的价值。

表 3.1 虚拟化管理应用程序（包括命令行工具）

应用程序	描述
virt-manager	虚拟机桌面管理工具
virt-install	虚拟机配给工具
virt-clone	虚拟机映像克隆工具
virt-image	从一个 XML 描述符构造虚拟机
virt-viewer	虚拟机图形控制台
virsh	virsh Guest 域的交互式终端

virt-manager 由 Red Hat 使用 Python 语言开发，用于控制虚拟机的生命周期，包括配给、虚拟网络管理，统计数据收集和报告，以及提供对虚拟机本身的简单图形访问。virt-manager 使用 libvirt 虚拟化库来管理可用的虚拟机管理程序。libvirt 公开了一个应用程序编程接口（API），该接口与大量开源虚拟机管理程序相集成，以实现控制和监视。libvirt 提供了一个名为 libvirtd 的守护程序，帮助实施控制和监视，如图 3.10 的一个简单堆栈中所示。

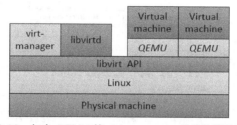

图 3.10 包含 QEMU 的 virt-manager 堆栈的简单表示

3.3.3 任务实施

步骤 1

打开应用程序—系统工具—虚拟化管理器，或者直接运行 virt-manager，如图 3.11 所示。

图 3.11 打开虚拟化管理器

步骤 2

打开虚拟化管理器窗口，如图 3.12 所示。

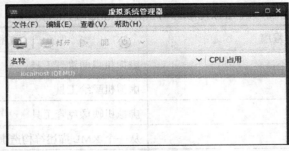

图 3.12　虚拟化管理器界面

步骤 3

单击"创建虚拟机"图标，输入虚拟机名称，如图 3.13 所示。

图 3.13　虚拟机创建对话框

步骤 4

选择 ISO 映像，输入映像文件路径，如图 3.14 所示。

图 3.14　新建虚拟机映像选择

步骤 5

根据实际情况输入虚拟内存大小和虚拟 CPU 个数,如图 3.15 所示。

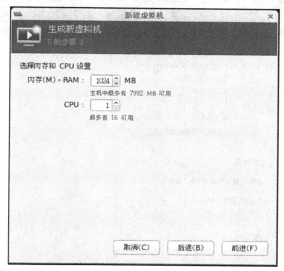

图 3.15　设置虚拟内存大小和虚拟 CPU 个数

步骤 6

按实际需要分配磁盘空间,如图 3.16 所示。

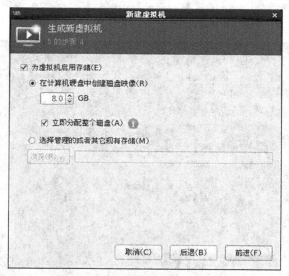

图 3.16　分配磁盘空间

步骤 7

主机设备选择桥接,如图 3.17 所示。

步骤 8

完成后,选择打开虚拟机,开始安装系统,如图 3.18 所示。

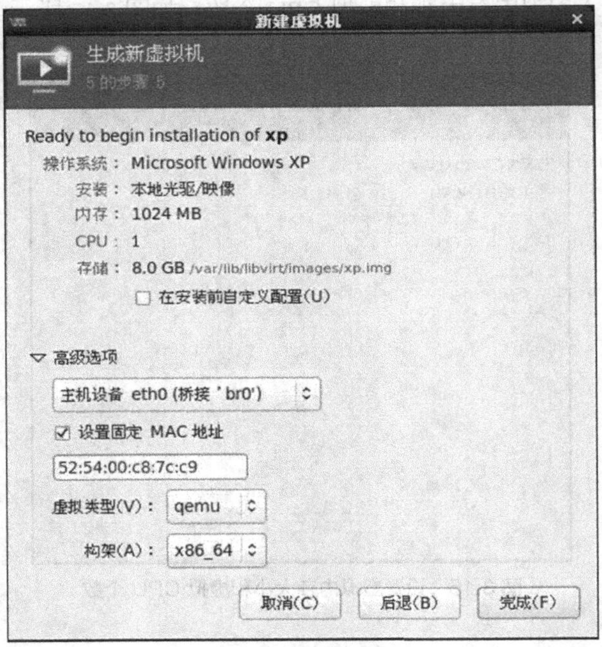

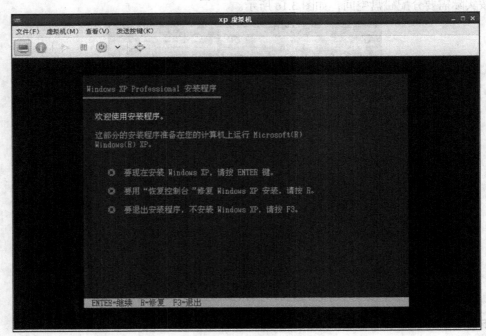

图 3.17 选择桥接

图 3.18 开始安装系统

步骤 9

虚拟机安装完成后，打开 IE 浏览器浏览网页，如图 3.19 所示。

图 3.19　虚拟机浏览网页

3.4　任务四　虚拟机的运行与远程访问

3.4.1　任务描述

虚拟机的运行和远程访问是必须掌握的技术，本节任务主要完成虚拟机的远程访问和设置，使用 Remote Viewer、VNC、SPICE 客户端工具连接控制台。通过本节的学习，读者将掌握虚拟机的远程访问配置，了解 VNC 协议和 SPICE 协议。

3.4.2　相关知识

1．virt-viewer

virt-viewer 是用于显示虚拟机的图形控制台最小的工具。控制台使用 VNC 或 SPICE 协议。客户机可以是基于名称、ID 或 UUID。如果客户机没有运行，那么 virt-viewer 试图等待控制台的连接，virt-viewer 通过连接远程主机查找控制台信息，然后通过网络传输连接到远程控制台。

2．VNC

Tiger-vnc 是一款优秀的 VNC 软件。VNC（Virtual Network Computing），为一种使用 RFB 协议的屏幕画面分享及远程操作软件。此软件借由网络，可发送键盘与鼠标的动作及实时的屏幕画面。VNC 与操作系统无关，因此可跨平台使用，如可用 Windows 连接到某 Linux 的电脑，反之亦可。甚至在没有安装客户端程序的电脑中，只要有支持 Java 的浏览器，也可使用。

VNC 系统由客户端、服务端和一个协议组成，VNC 的服务端目的是分享其所运行机器的屏幕，服务端被动地允许客户端控制它。VNC 客户端（或 Viewer）观察控制服务端，与服务端交互。VNC 协议 Protocol（RFB）是一个简单的协议，传送服务端的原始图像到客户端

(一个 X,Y 位置上的正方形的点阵数据），客户端传送事件消息到服务端。

Linux 上的 VNC 称为 xvnc，同时扮演两种角色，对 X 窗口系统的应用程序来说它是 X server，对于 VNC 客户端来说它是 VNC 伺服程序。

VNC 并非是安全的协议，虽然 VNC 伺服程序需设置密码才可接受外来连接，且 VNC 客户端与 VNC 伺服程序之间的密码传输经过加密，但仍可被轻易地拦截到并使用暴力搜索法破解。不过 VNC 可设计以 SSH 或 VPN 传输，以增加安全性。

3．SPICE

SPICE（独立计算环境简单协议）是红帽企业虚拟化桌面版的三大主要技术组件之一，具有自适应能力的远程提交协议，能够提供与物理桌面完全相同的最终用户体验。它包含有 3 个组件：

SPICE Driver——SPICE 驱动器，存在于每个虚拟桌面内的组件；

SPICE Device——SPICE 设备，存在于红帽企业虚拟化 Hypervisor 内的组件；

SPICE Client——SPICE 客户端，存在于终端设备上的组件，可以是瘦客户机或专用的 PC，用于接入每个虚拟桌面。

这 3 个组件协作运行，确定处理图形的最高效位置，以能够最大程度改善用户体验并降低系统负荷。如果客户机足够强大，SPICE 向客户机发送图形命令，并在客户机中对图形进行处理，显著减轻服务器的负荷。如果客户机不够强大，SPICE 在主机处理图形，从 CPU 的角度讲，图形处理并不需要太多费用。

3.4.3 任务实施

步骤 1

查看虚拟机的基本信息，如图 3.20 所示。

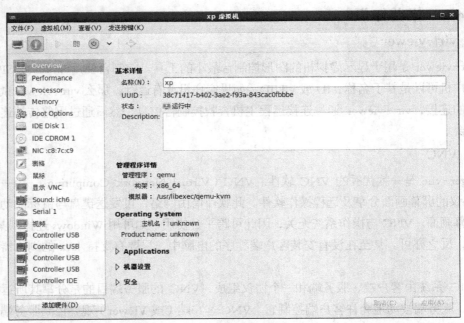

图 3.20 查看虚拟机的基本信息

步骤 2

显示有两种类型：VNC 和 SPICE，默认使用 VNC，地址是 127.0.0.1，端口号自动分配。如图 3.21 所示。

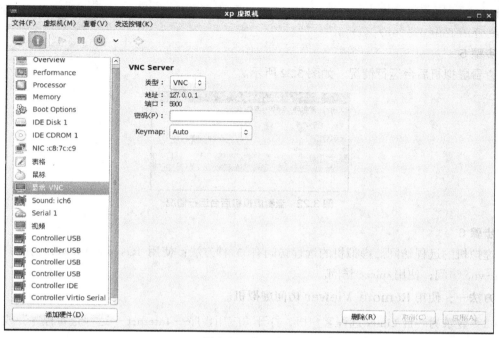

图 3.21　显示类型

步骤 3

修改侦听端口，实现远程访问。

默认侦听地址改为 0.0.0.0，允许远程访问，编辑文件 /etc/libvirt/qemu.conf。操作命令如下。

[root@localhost ~]# vi /etc/libvirt/qemu.conf

修改两处，删除左边"#"号。如下所示。

vnc_listen = "0.0.0.0"

spice_listen = "0.0.0.0"

也可以修改默认密码，删除左边"#"号。如下所示。

vnc_password = "XYZ12345"

spice_password = "XYZ12345"

修改后重启服务，如下所示。

[root@localhost ~]# service libvirtd restart
正在关闭 libvirtd 守护进程：　　　　　　　　　　　　　　　　　[确定]
启动 libvirtd 守护进程：　　　　　　　　　　　　　　　　　　　[确定]

步骤 4

查看进程情况，如下所示。

[root@localhost ~]# netstat -anp|grep 5900
tcp   0   127.0.0.1:5900   0.0.0.0:*

```
LISTEN           8540/qemu-kvm
tcp        0        0 127.0.0.1:5900           127.0.0.1:54468
ESTABLISHED 8540/qemu-kvm
tcp        0        0 127.0.0.1:54468          127.0.0.1:5900
ESTABLISHED 8788/python
```

步骤 5

查看虚拟机后台运行情况，如图 3.22 所示。

图 3.22　查看虚拟机后台运行情况

步骤 6

虚拟机的远程访问。虚拟机的远程访问有 3 种方法：使用 Remote Viewer 访问；使用 Tiger-vnc 访问；使用 spicec 访问。

方法一：使用 Remote Viewer 访问虚拟机。

（1）登录另一台 Linux 远程客户机，打开"应用程序—Internet—远程查看程序"或直接运行 Remote Viewer，如图 3.23 所示。

图 3.23　打开 RemoteViewer

如果使用 VNC，则 URL 输入 "vnc://IP 地址:端口号"；如果使用 SPICE，则 URL 输入 "spice://IP 地址:端口号"，如图 3.24 所示。

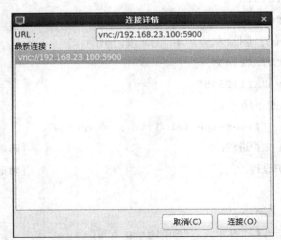

图 3.24　输入"VNC://IP 地址:端口号"

（2）单击"连接"按钮，如图 3.25 所示。

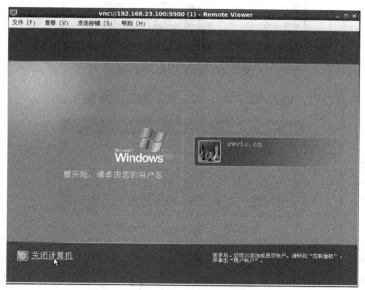

图 3.25 单击"连接"按钮显示信息

如果使用 Windows 系统远程登录，需要安装 virt-viewer 软件，下载地址：
https://fedorahosted.org/released/virt-viewer/virt-viewer-x64-2.0.msi
https://fedorahosted.org/released/virt-viewer/virt-viewer-x86-2.0.msi
安装后运行 Remote Viewer，如图 3.26 所示。

图 3.26 运行 Remote Viewer

方法二：使用 Tiger-vnc 访问虚拟机。

（1）登录另一台 Linux 远程客户机，打开"应用程序—Internet—远程查看程序"或直接运行 VNC Viewer，如图 3.27 所示。

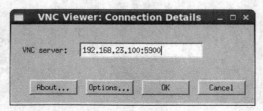

图 3.27 运行 VNC Viewer

(2)打开对话窗口后,输入 IP 地址:端口号,如图 3.28 所示。

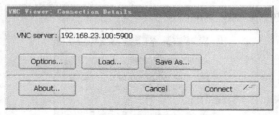

图 3.28 VNC viewer 对话窗口

如果使用 Windows 系统远程登录,需要安装 Tiger-vnc 软件。下载地址:

http://liquidtelecom.dl.sourceforge.net/project/tigervnc/tigervnc/1.3.1/vncviewer64.exe,

http://liquidtelecom.dl.sourceforge.net/project/tigervnc/tigervnc/1.3.1/vncviewer.exe。

(3)打开 VNC Viewer 运行窗口,输入 IP 地址:端口号,如图 3.29 所示。

图 3.29 VNC Viewer 运行窗口

(4)登录 Windows,使用 VNC Viewer 打开虚拟机,如图 3.30 所示。

方法三:使用 SPICEC 访问虚拟机。

(1)登录另一台 Linux 远程客户机,运行 SPICEC。操作命令如下。

[root@localhost ~]#spicec -h 192.168.23.100 -p 5900

spicec 命令连接:

spicec-h IPaddress -p port -w password

其中:-h 参数是 kvm 虚拟机 ip 地址,-p 参数是 kvm 虚拟机端口,-w 参数是密码。
如果使用 Windows 系统远程登录,需要安装 spice-client 软件。下载地址:

http://www.spice-space.org/download/binaries/spice-client-win32-0.6.3.zip

http://www.spice-space.org/download/stable/wspice-x86_20110308.zip

(2)spice-client 运行后,输入 IP 地址和端口号,单击 Connect 按钮,如图 3.31 所示。

图 3.30　使用 VNC Viewer 打开虚拟机

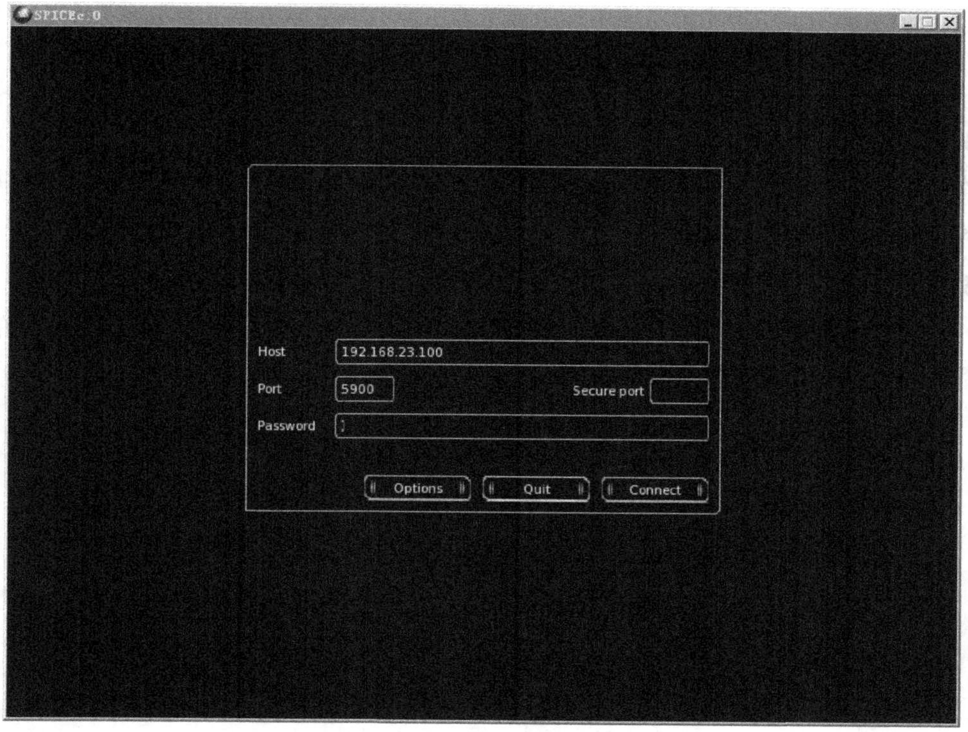

图 3.31　输入 IP 地址和端口号

（3）登录虚拟机，如图 3.32 所示。

图 3.32 登录虚拟机

3.5 任务五 使用 virsh-install 安装虚拟机

3.5.1 任务描述

本节任务主要是了解命令 virsh-install 的参数,掌握命令 virsh-install 参数的使用,掌握命令 virsh-install 的使用。

3.5.2 相关知识

virsh-install 是安装虚拟机的命令,方便用户在命令窗口上安装虚拟机,命令包含许多配置参数,virsh-install 的几个主要参数如下。

```
-h, --help                    #显示帮助信息
-n NAME, --name=NAME          #客户端事件名称
-r MEMORY, --ram=MEMORY       #以 MB 为单位为客户端事件分配的内存
--vcpus=VCPUS                 #虚拟机 CPU 数量,配置如:
    #--vcpus 5
    #--vcpus 5,maxcpus=10
    #--vcpus sockets=2,cores=4,threads=2
--disk path,size              #指定虚拟机镜像,size 指定分配大小单位为 G
```

```
--network bridge=br0              #网络类型，一般使用桥接
--network network:default         #默认使用NAT模式
--accelerate                      #表示使用内核加速功能
-c CDROM, --cdrom=CDROM           #光驱安装介质
-l LOCATION, --location=LOCATION
    #安装源(例如: nfs:host:/path、http://host/path
    #ftp://host/path)
--vnc                             #启用vnc远程管理
--vncport                         #指定vnc监控端口，默认端口为5900，端口不能重复
--vnclisten                       #指定vnc绑定IP，默认绑定127.0.0.1，改为0.0.0.0
```

3.5.3 任务实施

步骤1

打开控制台，输入virsh-install命令。如下所示。

```
[root@localhost ~]# virt-install --name=CentOS --ram 1024 --vcpus=2 --disk \
path=/var/lib/libvirt/images/CentOS.img,size=8,bus=virtio --accelerate \
--cdrom /dev/cdrom  --vnc --vncport=5910 --vnclisten=0.0.0.0 \
--network bridge=br0,model=virtio -noautoconsole

WARNING  KVM acceleration not available, using 'qemu'


开始安装......
正在分配 'CentOS.img'                          | 8.0 GB     00:00
创建域......                                    |   0 B      00:00
域安装仍在进行。您可以重新连接
到控制台以便完成安装进程
```

步骤2

使用virsh-install命令只完成虚拟机的创建，虚拟机的安装还需要在远程机器控制台完成安装，本实训使用TigerVNC Viewer打开虚拟机进行CentOS系统安装，如图3.33所示。

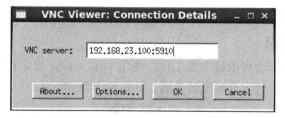

图3.33 TigerVNC Viewer

步骤3

进入操作界面进行虚拟机的安装，如图3.34所示。

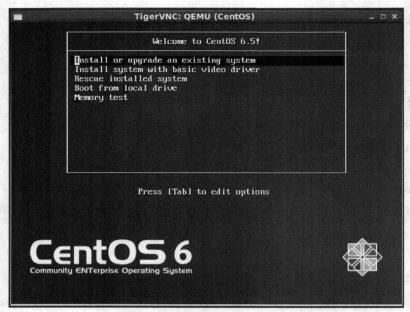

图 3.34 虚拟机的安装界面

如果安装 Windows 系列虚拟机，出现找不到硬盘，可以删除 virtio 选项，另外安装盘是 iso 文件，需要修改 qemu.conf 配置文件。操作命令：

[root@localhost ~]# vi /etc/libvirt/qemu.conf

把下面几个地方的注释去掉，把 dynamic_ownership=1 改为 0，

部分内容修改为

user="root"

group="root"

dynamic_ownership=0

3.6 任务六 使用 virsh-clone 克隆虚拟机

3.6.1 任务描述

本节任务主要是了解命令 virsh-clone 的参数，掌握命令 virsh-clone 参数的使用，掌握命令 virsh-clone 克隆虚拟机。

3.6.2 相关知识

克隆虚拟机在虚拟化管理中是常见工作，使用 virsh-clone 命令克隆虚拟机可以方便我们的工作。virsh-clone 命令包含许多配置参数，其主要参数如下。

```
-h, --help                          #显示帮助
--connect=URI                       #连接接到 hypervisorI
-o ORIGINAL_GUEST, --original=ORIGINAL_GUEST
                                    #原始客体的名称；必须为关闭或暂停状态
--original-xml=ORIGINAL_XML         #将 XML 文件作为原始客体使用
```

```
--auto-clone                              #从初始客体配置中自动生成克隆名称和存储路径
-n NEW_NAME, --name=NEW_NAME              #新客户端的名称
-u NEW_UUID, --uuid=NEW_UUID              #克隆客户端的新 UUID；默认为随机生成 UUID
-f NEW_DISKFILE, --file=NEW_DISKFILE      #作为新客户端磁盘映像的新文件
```

3.6.3 任务实施

步骤 1

打开控制台，输入 virsh-clone 命令。如下所示。

```
[root@localhost ~]# virt-clone --connect qemu:///system --original=xp --name=xp2 --file=/var/lib/libvirt/images/xp2.img
ERROR    必须暂停或者关闭有要克隆设备的域
```

错误信息表示克隆时需要暂停虚拟机或者关闭，所以需要暂停虚拟机。关闭命令：

```
[root@localhost ~]# virsh suspend xp
域 xp 被挂起
```

重新克隆，如下所示。

```
[root@localhost ~]# virt-clone --connect qemu:///system --original=xp --name=xp2 --file=/var/lib/libvirt/images/xp2.img
正在分配 'xp2.img'                                      | 8.0 GB    01:27

Clone 'xp2' created successfully.
```

克隆成功后生成的文件如下。

```
[root@localhost ~]# cd /etc/libvirt/qemu
[root@localhost qemu]# ls
CentOS.xml  networks  xp2.xml  xp.xml
[root@localhost qemu]# ls /var/lib/libvirt/images/
CentOS.img  xp2.img  xp.img
```

步骤 2

修改配置，克隆出来的 xp2 需要修改端口，避免端口冲突。运行 virsh 管理工具，编辑 xp2 配置文件。如下所示。

```
[root@localhost qemu]# virsh
欢迎使用 virsh，虚拟化的交互式终端


输入： 'help' 来获得命令的帮助信息
      'quit' 退出


virsh # list --all
 Id    名称                           状态
----------------------------------------------------
```

```
 11    xp                              暂停
 16    CentOS                          running
  -    xp2                             关闭
```

```
virsh # edit xp2
```

找到如下内容。

```
<graphics type='vnc' port='-1' autoport='yes' listen='0.0.0.0'>
<listen type='address' address='0.0.0.0'/>
</graphics>
```

将其修改为如下内容。

```
<graphics type='vnc' port='5902' autoport='no' listen='0.0.0.0'>
<listen type='address' address='0.0.0.0'/>
</graphics>
```

重启虚拟机。操作命令:

```
[root@localhost ~]# virsh start xp2
```

域 xp2 已开始

步骤 3

查看虚拟机运行状态，操作命令:

```
[root@localhost ~]# virsh list
 Id    名称                              状态
----------------------------------------------------
 11    xp                              暂停
 16    CentOS                          running
 17    xp2                             running
```

修改配置也可以直接使用 vi 编辑，操作命令:

```
[root@localhost ~]# vi /etc/libvirt/qemu/xp2.xml
[root@localhost ~]# service libvirtd restart
```

步骤 4

打开远程控制台，使用 Remote Viewer 或 VNC 或 SPICE 客户端工具打开克隆虚拟机。

3.7 任务七 virsh 命令的使用

3.7.1 任务描述

本节任务主要是掌握 virsh 命令的使用,掌握 qemu 命令的使用。

3.7.2 相关知识

1. virsh

virsh（Virtual Shell）是由一个名为 libvirt 的软件提供的管理工具，提供管理虚拟机更高级的能力。virsh 大部分的功能与 xm 一样，可以用来启动、删除、控制、监控 Xen 的区域，因此也可以用来管理 Xen 中所有的虚拟机。

2．QEMU

QEMU 是一套由 Fabrice Bellard 所编写的以 GPL 许可证分发源码的模拟处理器，在 GNU/Linux 平台上使用广泛。Bochs，PearPC 等与其类似，但不具备其许多特性，比如高速度及跨平台的特性，通过 KQEMU 这个闭源的加速器，QEMU 能模拟至接近真实电脑的速度。

QEMU 支持多种架构，可以模拟 IA-32（x86）个人电脑、AMD 64 位个人电脑、MIPS R4000、升阳的 SPARCsun3 与 PowerPC（PReP 及 Power Macintosh）架构。

3.7.3 任务实施

步骤 1

练习以下几个命令。

```
virsh list                          #列出正在运行的虚拟机
virsh list --all                    #列出所有的虚拟机
virsh autostart xp                  #自动开始 xp 域
virsh start xp                      #开始启动 xp 域
virsh suspend xp                    #挂起，暂停 xp 域
virsh resume xp                     #重新恢复 xp 域
virsh undefine xp                   #删除 xp 域
virsh destroy xp                    #强制 xp 域关机
virsh shutdown xp                   #执行 xp 域关机操作
virsh reboot xp                     #重启 xp 域
virsh dominfo xp                    #域的基本信息
virsh nodeinfo                      #物理机的信息
virsh dommemstat xp                 #域的内存信息
virsh setmem xp                     #设置内存大小默认单位是 kb
virsh vncdisplay xp                 #vnc 连接的 ip 地址和端口
virsh edit xp                       #编辑 xp 域的 XML 文件
virsh snapshot-create-as xp xpshot  #快照
virsh snapshot-list xp              #列出快照
virsh snapshot-delete xp xpshot     #删除快照
virsh snapshot-revert xp xpshot     #恢复快照
```

步骤 2

练习下面几种常见操作。

（1）修改参数。

```
virsh edit xp               #注意 vi 直接编辑/etc/libvirt/qemu/xp.xml 不生效
virsh list --all            #显示运行状态下的虚拟机
virsh shutdown xp           #先关闭虚拟机
```

```
virsh start xp                    #再启动虚拟机
```

（2）强制关掉宿主机导致宿主机开机后不能启动 xp。

```
virsh undefine xp
virsh managedsave-remove xp
virsh start xp
```

（3）修改域名。

```
cp xp.xml myxp.xml
virsh undefine xp
vi myxp.xml
```

修改的内容如下：

```
<domain type='kvm'>
<name>myxp</name># 修改域名


virsh define myxp.xml
virsh list -all
```

（4）删除域。

```
virsh undefine xp                 #正在工作也能删除
virsh destroy xp
```

（5）启动虚拟机的两种方式。

```
virsh start xp
virsh create /etc/libvirt/qemu/xp.xml
```

（6）备份。

```
virsh dumpxml nodeA >nodeAback.xml
virsh save nodeA nodeA_bak.img
virsh restore nodeA_bak.img
```

步骤 3

练习使用下面命令。

（1）使用 qemu-img 创建磁盘文件，格式如下。

```
qemu-img create [-f fmt] [-o options] filename [size]
```

作用：创建一个格式为 fmt，大小为 size，文件名为 filename 的镜像文件。例如：

```
qemu-img create -f qcow2 xp.qcow2 8G
```

（2）使用 qemu-img 创建快照，格式如下。

```
qemu-img snapshot [-l | -a snapshot | -c snapshot | -d snapshot] filename
```

其中：-l 查询并列出镜像文件所有快照。

-a 使用某个快照

-c 创建一个快照

-d 删除一个快照

例如：

```
qemu-img snapshot -c xpsnap1 xp.qcow2
```

（3）使用 qemu-img 转换镜像文件格式，格式如下。

```
qemu-img convert [-c] [-f fmt] [-O output_fmt] [-o options] filename  output_filename
```

作用：将 fmt 格式的 filename 镜像文件根据 options 选项转换为格式为 output_fmt 的名为 output_filename 的镜像文件。例如：

```
qemu-img convert -O qcow2 xp.img xp-a.img
```

（4）使用 qemu-kvm 创建虚拟机。例如：

```
qemu-kvm -m 1024 -localtime -M pc -smp 1 -drive file=win7.img,cache=writeback,boot=on
 -net nic,macaddr=00:0C:29:12:34:80 -net tap -cdrom Windows7.iso \
 -boot d -name kvm-win7,process=kvm-win7 -vnc :2 -usb -usbdevice tablet
```

说明：

```
  -m 1024                              #设置虚拟系统内存1024MB
  -localtime                           #使虚拟系统与宿主系统时间一致
-M pc                                  #虚拟系统类型为pc
  -smp 1                               #1个CPU
-drive file=win7.img,cache=writeback,boot=on #硬盘选项，虚拟磁盘是win7.img
#cache方式为writeback，可引导型磁盘。
  -net nic,macaddr=00:0C:29:12:34:80   #网卡选项，手工指定mac地址
  -net tap                             #tap类型网络，相当于"桥模式"
  -cdrom Windows7.iso                  #光驱
  -boot d                              #启动顺序，d代表光驱
  -name kvm-win7,process=kvm-win7      #为虚拟机取名，便于识别
  -vnc :2                              #这里是通过vnc连接控制窗口，
#这里是在5902端口。client可用IP:2连接
  -usb -usbdevice tablet               #启用usb设备中的tablet功能
#开启该功能可使虚拟机内外的鼠标同步
```

另外，在安装了磁盘和网卡的半虚拟化驱动后，可以在-drive 中加入 if=virtio 使用磁盘半虚拟化，在-net nic 中加入 model=virtio 使用网卡半虚拟化驱动。

3.8 本章小结

本章主要任务是完成虚拟化的配置和使用，学习 virt-viewer、SPICE Client、Tiger-vnc 远程客户端工具的使用，学习使用 virsh-install 安装虚拟机，学习使用 virsh-clone 克隆虚拟机。通过本章的学习，能加深读者对虚拟化技术的理解和使用。

第 4 章 MPI——面向计算的集群技术

随着虚拟化技术和云计算技术的发展，越来越多的科学计算应用运行在云计算资源之上。MPI 编程模型是一种消息传递编程模型，大多数科学计算应用都是基于这种消息传递编程模型的高性能计算应用，其对网络 I/O 负载较为敏感。MPI 应用是一类广泛应用的由多个进程协同工作的并行计算应用，在云计算环境下，其进程运行在多个不同的虚拟机之中。MPI 是集群之间通信机制的一种重要模型，通过本章学习 MPI，我们正式进入并行计算环境，这也是云计算的真正开始，熟悉在并行环境下的工作方式将有助于我们理解云计算技术和云计算集群的通信机制。MPICH 是 MPI 最流行的非专利实现，其版本基本与 MPI 标准基本同步。MPICH 并行环境的建立主要完成以下 3 项工作。

（1）配置好 NFS 服务，实现所有节点对主节点指定文件夹的共享，该文件夹为 MPICH 的安装位置、数据和程序的存储位置，这样就可以避免在每个节点安装 MPICH，启动计算时也可以避免每次向各个节点分发程序。

（2）配置好节点间的互信，这一步就是实现集群内部各节点间无需密码访问，因为 MPICH 在计算时需要在各节点进行数据交换，集群内的节点应用相互信任的节点。

（3）编译安装配置 MPICH。

4.1 任务一　配置 ssh 实现节点间无密码访问

4.1.1 任务描述

由于 MPI 并行程序需要在各节点间进行信息传递，所以必须实现所有节点两两之间能无密码访问。节点间的无密码访问是通过配置 ssh 证书认证来实现的。配置 ssh 是集群系统配置的常用操作，MPI、Hadoop 等系统均需配置 ssh。本节任务是配置 ssh 实现 MPI 节点间无密码访问。通过配置 ssh 我们将了解非对称加密算法，了解数字证书的作用，完成 ssh 无密码访问配置和使用。

4.1.2 相关知识

1．SSH 的概念和作用

SSH 为 Secure Shell 的缩写，由 IETF 的网络工作小组（Network Working Group）所制定，

是建立在应用层和传输层基础上的安全协议。SSH 是一个用来替代 TELNET、FTP 以及 R 命令的工具包，主要是想解决口令在网上明文传输的问题。SSH 是目前较可靠、专为远程登录会话和其他网络服务提供安全性的协议。利用 SSH 协议可以有效防止远程管理过程中的信息泄露问题。SSH 最初是 UNIX 系统上的一个程序，后来又迅速扩展到其他操作平台。SSH 在正确使用时可弥补网络中的漏洞。SSH 客户端适用于多种平台。几乎所有 UNIX 平台包括 HP-UX、Linux、AIX、Solaris、Digital UNIX、Irix，都可运行 SSH。为了系统安全和用户自身的权益，推广 SSH 是必要的。

传统的网络服务程序，如：ftp、pop 和 telnet 在本质上都是不安全的，因为它们在网络上用明文传送口令和数据，别有用心的人非常容易截获这些口令和数据。而且，这些服务程序的安全验证方式也是有其弱点的，就是很容易受到"中间人"（Man-in-the-Middle）这种方式的攻击。所谓"中间人"的攻击方式，就是"中间人"冒充真正的服务器接收你传给服务器的数据，然后再冒充你把数据传给真正的服务器。服务器和你之间的数据传送被"中间人"一转手做了手脚之后，就会出现很严重的问题。通过使用 SSH，你可以把所有传输的数据进行加密，这样"中间人"这种攻击方式就不可能实现了，而且也能够防止 DNS 欺骗和 IP 欺骗。使用 SSH，还有一个额外的好处就是传输的数据是经过压缩的，所以可以加快传输的速度。SSH 有很多功能，它既可以代替 Telnet，又可以为 FTP、PoP，甚至为 PPP 提供一个安全的"通道"。

2. SSH 的工作原理

SSH 是由客户端和服务端的软件组成的。服务端是一个守护进程，他在后台运行并响应来自客户端的连接请求。服务端一般是 sshd 进程，提供了对远程连接的处理，一般包括公共密钥认证、密钥交换、对称密钥加密和非安全连接。客户端包含 ssh 程序以及像 scp（远程复制）、slogin（远程登录）、sftp（安全文件传输）等其他的应用程序。他们的工作机制大致是本地的客户端发送一个连接请求到远程的服务端，服务端检查申请的包和 IP 地址再发送密钥给 SSH 的客户端，本地再将密钥发回给服务端，自此连接建立。一旦建立一个安全传输层连接，客户机就发送一个服务请求。当用户认证完成之后，会发送第二个服务请求。这样就允许新定义的协议可以与上述协议共存。

SSH 主要由 3 部分组成。

（1）传输层协议（[SSH-TRANS]）。

它提供了服务器认证，保密性及完整性。此外它有时还提供压缩功能。SSH-TRANS 通常运行在 TCP/IP 连接上，也可能用于其他可靠数据流上。SSH-TRANS 提供了强力的加密技术、密码主机认证及完整性保护。该协议中的认证基于主机，并且该协议不执行用户认证。更高层的用户认证协议可以设计为在此协议之上。

（2）用户认证协议（SSH-USERAUTH）。

用于向服务器提供客户端用户鉴别功能。它运行在传输层协议 SSH-TRANS 上面。当 SSH-USERAUTH 开始后，它从低层协议那里接收会话标识符（从第一次密钥交换中的交换哈希 H）。会话标识符唯一标识此会话并且适用于标记以证明私钥的所有权。SSH-USERAUTH 也需要知道低层协议是否提供保密性保护。

(3) 连接协议 (SSH-CONNECT)。

将多个加密隧道分成逻辑通道。它运行在用户认证协议上。它提供了交互式登录话路、远程命令执行、转发 TCP/IP 连接和转发 X11 连接。

由于 SSH 的源代码是公开的,所以在 UNIX 世界里它获得了广泛的认可。Linux 的源代码也是公开的,大众可以免费获得,并同时获得了类似的认可。这就使得所有开发者(或任何人)都可以通过补丁程序或 bug 修补来提高其性能,甚至还可以增加功能。获得并安装 SSH 意味着其性能可以不断得到提高而无须得到来自创作者的直接技术支持。SSH 替代了不安全的远程应用程序。通过使用 SSH, 即使在不安全的网络中发送信息时不必担心会被监听。用户也可以使用 POP 通道和 Telnet 方式, 通过 SSH 可以利用 PPP 通道创建一个虚拟个人网络 (Virtual Private Network, V P N)。

4.1.3 任务实施

本章搭建由四台节点机组成集群环境,每个节点机上安装 CentOS-6.5-x86_64 系统。四台节点机使用的 IP 地址分别为 192.168.23.111、192.168.23.112、192.168.23.113、192.168.23.114, 对应节点主机名为 node1、node2、node3、node4。

步骤 1

创建用户, 在四台节点机分别创建用户 mpi, uid=600, 设置 mpi 用户无密码相互访问, 密码分别为 mpi1111, mpi2222, mpi3333, mpi4444。操作命令:

```
[root@node1 ~]# useradd -u 600 mpi
[root@node1 ~]# passwd mpi


[root@node2 ~]# useradd -u 600 mpi
[root@node2 ~]# passwd mpi


[root@node3 ~]# useradd -u 600 mpi
[root@node3 ~]# passwd mpi


[root@node4 ~]# useradd -u 600 mpi
[root@node4 ~]# passwd mpi
```

步骤 2

生成证书, 打开 node1 节点机, 以用户 mpi 登录。操作命令:

```
Localhost login: mpi
Password:
[mpi@node1 ~]$ _
```

或者以 root 登录后使用命令 su 切换到 mpi 用户。操作命令:

```
[root@node1 ~]# su - mpi
[mpi@node1 ~]$ _
```

步骤 3

使用 ssh-keygen 生成证书密钥，证书密钥有两种算法：dsa 和 rsa。操作命令：

```
[mpi@node1 ~]$ ssh-keygen -t dsa
```

执行后，显示如下信息，按回车就可以，如下所示。

```
Generating public/private dsa key pair.
Enter file in which to save the key (/home/mpi/.ssh/id_dsa):
Created directory '/home/mpi/.ssh'.
Enter passphrase (empty for no passphrase):
Enter same passphrase again:
Your identification has been saved in /home/mpi/.ssh/id_dsa.
Your public key has been saved in /home/mpi/.ssh/id_dsa.pub.
The key fingerprint is:
91:d8:63:08:0d:22:13:a4:93:18:20:cb:cb:e4:74:01 mpi@node1
The key's randomart image is:
+--[ DSA 1024]----+
|XEo.oo           |
|=* ....+ .       |
|*+ . o *         |
|=.o   . o        |
| +     S         |
|                 |
|                 |
|                 |
|                 |
+-----------------+
```

步骤 4

使用 ssh-copy-id 分别复制证书公钥到 node1，node2，node3，node4 节点机上。操作命令：

```
[mpi@node1 ~]$ ssh-copy-id -i .ssh/id_dsa.pub mpi@node1
[mpi@node1 ~]$ ssh-copy-id -i .ssh/id_dsa.pub mpi@node2
[mpi@node1 ~]$ ssh-copy-id -i .ssh/id_dsa.pub mpi@node3
[mpi@node1 ~]$ ssh-copy-id -i .ssh/id_dsa.pub mpi@node4
```

第一次复制需要先输入"yes"，再输入密码。下面是公钥到节点机 node1 的提示信息。

```
The authenticicfy of host 'node1 (192.168.23.111)' can;t be established.
RSA key fingerprint is 20:ea:5f:1c:37:90:71:4b:13:5c:92:f1:ee:56:9e:f4.
Are you sure want to continue connecting (yes/no)? yes
Warning: Permanently added 'node1,192.168.23.111' (RSA) to the list of known
 hosts
mpi@node1's password:
```

```
Now try logging into the machine, with "ssh 'mpi@node1'", and check in:

  .ssh/authorized_keys

to make sure we haven't added extra keys that you weren't expecting.
```

步骤 5

分别使用 ssh 登录 node1、node2、node3、node4 测试,测试 ssh 无密码登录 node1 和 node2。操作命令:

```
[mpi@node1 ~]$ ssh node1              [mpi@node1 ~]$ ssh node2
[mpi@node1 ~]$ exit                   [mpi@node1 ~]$ exit
Logout                                logout
Connection to node1 closed.           Connection to node2 closed.
```

步骤 6

使用 scp 分别复制证书私钥到 node2、node3、node4 节点机上,实现相互无密码登录。操作命令:

```
[mpi@node1 ~]$ scp .ssh/id_dsa node2:/home/mpi/.ssh
[mpi@node1 ~]$ scp .ssh/id_dsa node3:/home/mpi/.ssh
[mpi@node1 ~]$ scp .ssh/id_dsa node4:/home/mpi/.ssh
```

步骤 7

测试相互登录,在 node2 节点机上测试 ssh 登录 node3 节点机,第一次需要输入"yes",其他测试相同。如下所示。

```
[mpi@node2~]$ ssh node3
The authenticify of host 'node3 (192.168.23.113)' can;t be established.
RSA key fingerprint is 20:ea:5f:1c:37:90:71:4b:13:5c:92:f1:ee:56:9e:f4.
Are you sure want to continue connecting (yes/no)? yes
Warning: Permanently added 'node3,192.168.23.113' (RSA) to the list of known hosts
Last login: Mon Dec 22 22:40:44 2014 from node2
```

第二次登录不需要输入"yes"。

```
[mpi@node2~]$ ssh node3
Last login: Mon Dec 22 23:15:45 2014 from node2
```

退出使用命令:exit 键、logout 键或按 Ctrl+D 组合键。

至此,ssh 无密码相互登录配置成功。

4.2 任务二 网络文件系统 NFS

4.2.1 任务描述

由于 MPICH 的安装目录和用户可执行程序在并行计算时需要在所有节点存副本,而且目录要相互对应，每次一个节点一个节点地复制非常麻烦，采用 NFS（Netword File Systom，网络文件系统）后可以实现所有节点内容与主节点内容同步更新，并自动实现目录的对应。本节任务主要是 NFS 安装配置与挂载，节点机 node1 设为 NFS 服务器，设置共享目录 /home/mpi，可读写，节点机 node2、node3、node4 为客户机，共享 NFS 服务。

4.2.2 相关知识

1. 什么是 NFS

NFS 是 Network File System 的缩写，即网络文件系统，是一种使用于分散式文件系统的协定，由 Sun 公司开发，于 1984 年向外公布。功能是通过网络让不同的机器、不同的操作系统能够彼此分享个别的数据，让应用程序在客户端通过网络访问位于服务器磁盘中的数据，是在类 Unix 系统间实现磁盘文件共享的一种方法。

NFS 的基本原则是"容许不同的客户端及服务端通过一组 RPC 分享相同的文件系统"，它是独立于操作系统，容许不同硬件及操作系统的系统共同进行文件的分享。NFS 在文件传送或信息传送过程中依赖于 RPC 协议。RPC，远程过程调用（Remote Procedure Call）是能使客户端执行其他系统中程序的一种机制。NFS 本身是没有提供信息传输的协议和功能的，但 NFS 却能让我们通过网络进行资料的分享，这是因为 NFS 使用了一些其他的传输协议，而这些传输协议用到了这个 RPC（远程过程调用协议）功能的。可以说 NFS 本身就是使用 RPC 的一个程序，或者说 NFS 也是一个 RPC Server。所以只要用到 NFS 的地方都要启动 RPC 服务，可以这么理解 RPC 和 NFS 的关系：NFS 是一个文件系统，而 RPC 负责信息的传输。

使用 NFS 能节省本地存储空间，将常用的数据存放在一台 NFS 服务器上且可以通过网络访问，那么本地终端将可以减少自身存储空间的使用。用户不需要在网络中的每个机器上都建有 Home 目录，可以将其放在 NFS 服务器上且可以在网络上被访问使用。一些存储设备如软驱、CD-ROM 和 Zip（一种高储存密度的磁盘驱动器与磁盘）等都可以在网络上被别的机器使用。这可以减少整个网络上可移动介质设备的数量。

2. 主服务端配置

配置 NFS 前要关闭所有节点的防火墙等安全设置，具体步骤参考第 2 章中的任务三。配置 NFS 前要开启主服务器端的 portmap 服务和 nfs 服务，并设置主服务器端的 /etc/exports 文件。/etc/exports 文件中的项的格式相当简单。要共享一个文件系统，只需要编辑 /etc/exports 并使用下面的格式给出这个文件系统（和选项）即可。

```
directory (or file system)    client1 (option1, option2) client2 (option1, option2)
```

常用选项有如下几种。

- secure：这个选项是缺省选项，它使用了 1024 以下的 TCP/IP 端口实现 NFS 的连接。指定 insecure 可以禁用这个选项。
- rw：这个选项允许 NFS 客户机进行读/写访问。缺省选项是只读的。
- async：这个选项可以改进性能，但是如果没有完全关闭 NFS 守护进程就重新启动了 NFS 服务器，这也可能会造成数据丢失。缺省设置为 sync。
- no_wdelay：这个选项关闭写延时。如果设置了 async，那么 NFS 就会忽略这个选项。

- nohide：如果将一个目录挂载到另外一个目录之上，那么原来的目录通常就被隐藏起来或看起来像空的一样。要禁用这种行为，需启用 hide 选项。
- no_subtree_check：这个选项关闭子树检查，子树检查会执行一些不想忽略的安全性检查。缺省选项是启用子树检查。
- no_auth_nlm：这个选项也可以作为 insecure_locks 指定，它告诉 NFS 守护进程不要对加锁请求进行认证。如果关心安全性问题，就要避免使用这个选项。缺省选项是 auth_nlm 或 secure_locks。
- mp(mountpoint=path)：通过显式地声明这个选项，NFS 要求挂载所导出的目录。
- fsid=num：这个选项通常都在 NFS 故障恢复的情况中使用。如果希望实现 NFS 的故障恢复，请参考 NFS 文档。

用户映射的选项包括以下几个。

- root_squash：这个选项不允许 root 用户访问挂载上来的 NFS 卷。
- no_root_squash：这个选项允许 root 用户访问挂载上来的 NFS 卷。
- all_squash：这个选项对于公共访问的 NFS 卷来说非常有用，它会限制所有的 UID 和 GID，只使用匿名用户。缺省设置是 no_all_squash。
- anonuid 和 anongid：这两个选项将匿名 UID 和 GID 修改成特定用户和组账号。

/etc/exports 的例子：

```
/home/mpi     192.168.23.*
/home/mpi     192.168.23.112
/home/mpi     192.168.23.112(rw, all_squash, anonuid=210, anongid=100)
/home/mpi     * (ro, insecure, all_squash)
```

第一项将 /home/mpi 目录导出给 192.168.23 网络中的所有主机。

第二项将 /home/mpi 导出给一台主机：192.168.23.112。

第三项指定了主机 192.168.0.112，并将对文件的读写权限授权给 user id=210 且 group id=100 的用户。

最后一项针对一个公共目录，它只有只读权限并只能允许以匿名账号的身份访问。

3. NFS 客户机

要使用 NFS 作为客户机，客户机机器必须要运行 rpc.statd 和 portmap 进程。如果它们正在运行（应该如此），那么就可以使用下面的通用命令来挂载服务器上导出的目录。

```
mount server:directory local mount point
```

通常来说，必须以 root 用户的身份来挂载文件系统。在远程计算机上，可以使用下面的命令（假设 NFS 服务器的 IP 地址是 192.168.23.111），命令如下。

```
mount 192.168.23.111:/home/mpi/home/mpi
```

您所使用的发行版可能会要求在挂载文件系统时指定文件系统的类型。如果出现这种情况，请执行下面的命令。

```
mount -t nfs 192.168.23.111:/home/mpi/home/mpi
```

如果服务器端已经正确设置好了，那么远程文件系统应该可以毫无问题地加载。现在，执行 cd 命令切换到客户端 /home/mpi 目录中，然后执行 ls 命令来查看文件。我们也可在所有子节点的 /etc/fstab 文件中输入以下的代码，使文件系统在启动时实现自动挂载 NFS（非必须）：

```
192.168.23.111:/home/mpi/home/mpi    nfs    defaults    0 0
```

4.2.3 任务实施

步骤 1

检查系统是否安装 nfs-utils 和 rpcbind 两软件包，下面表示系统已经安装了软件包。

```
[root@node1 ~]# rpm -qa|grep nfs-utils
nfs-utils-1.2.3-39.el6.x86_64
nfs-utils-lib-1.1.5-6.el6.x86_64
[root@node1 ~]# rpm -qa|grep rpcbind
rpcbind-0.2.0-11.el6.x86_64
```

如果没有安装，则安装命令如下。

```
[root@node1 ~]# cd /media/CentOS_6.5_Final/Packages/
[root@node1 Packages]# rpm -ivh nfs-utils-1.2.3-39.el6.x86_64.rpm
[root@node1 Packages]# rpm -ivh rpcbind-0.2.0-11.el6.x86_64.rpm
```

步骤 2

编辑配置文件"/etc/exports"，操作命令如下。

```
[root@node1 ~]# vi /etc/exports
```

编辑内容如下。

```
/home/mpi 192.168.23.0/24(rw,sync,no_all_squash)
```

rw 表示读写，sync 表示同步操作，no_all_squash 表示远程普通用户不映射到 nfsnobody。

步骤 3

启动 rpcbind 服务和 nfs 服务。如果服务已经启动，则不提示信息。操作命令如下。

```
[root@node1 ~]# service rpcbind start
[root@node1 ~]# service nfs start
```

如果服务原先没启动，则显示内容如下。

```
[root@node1 ~]# service rpcbind start
```
正在启动 rpcbind：   [确定]
```
[root@node1 ~]# service nfs start
```
启动 NFS 服务：   [确定]
关掉 NFS 配额：   [确定]
启动 NFS mountd：   [确定]
启动 NFS 守护进程：   [确定]
正在启动 RPC idmapd：   [确定]

步骤 4

使用 exportfs 命令，对 NFS 服务进行一些操作。

（1）输出所有目录，操作命令如下。

```
[root@node1 ~]# exportfs -av
 exporting 192.168.23.0/24:/home/mpi
```
（2）重新输出共享目录，操作命令如下。
```
[root@node1 ~]# exportfs -arv
 exporting 192.168.23.0/24:/home/mpi
```
（3）停止输出所有共享目录，操作命令如下。
```
[root@node1 ~]# exportfs -auv
```

步骤 5

查看 NFS 服务器共享。

（1）查看共享目录，操作命令如下。
```
[root@node1 ~]# showmount -e 192.168.23.111
Export list for 192.168.23.111:
/home/mpi 192.168.23.0/24
```
（2）查看客户机或 IP 地址及其在主机中的目录，操作命令如下。
```
[root@node1 ~]# showmount -a 192.168.23.111
All mount points on 192.168.23.111:
```

步骤 6

分别在 node2，node3，node4 节点机上挂载 nfs 共享目录，参数 "-t nfs" 可以省略。操作命令：
```
[root@node2 ~]# mount -t nfs 192.168.23.111:/home/mpi /home/mpi
[root@node3 ~]# mount 192.168.23.111:/home/mpi /home/mpi
[root@node4 ~]# mount 192.168.23.111:/home/mpi /home/mpi
```

步骤 7

测试 NFS 写操作。

（1）在 node4 节点机上写文件和列文件操作，操作命令如下。
```
[root@node4 ~]# su - mpi
[mpi@node4 ~]$ touch node4w1
[mpi@node4 ~]$ ll
总用量 0
-rw-rw-r--. 1 mpi mpi 0 12月 21 14:51 node4w1
```
（2）在 node1 节点机上检查文件是否存在，操作命令如下。
```
[root@node1 ~]# cd /home/mpi
[root@node1 mpi]# ll
总用量 0
-rw-rw-r--. 1 mpi mpi 0 12月 21 14:51 node4w1
```

至此，NFS 共享设置成功。

4.3 任务三 MPICH 编译运行

4.3.1 任务描述

MPICH 是 MPI（Message-Passing Interface）的一个应用实现，是用于并行运算的工具。因特网提供开源代码下载，需要相应的编译包编译，需要安装 GNU 编译器套件 GCC（GNU Compiler Collection）。GCC 包括 C、C++、Objective-C、Fortran、Java、Ada 和 Go 语言的前端，也包括了这些语言的库（如 libstdc++、libgcj 等）。本节任务主要介绍开发包的安装、mpich 的编译、安装和运行，使用的 mpich 为 mpich-3.1.3（stable release）版本。通过本节的学习掌握 mpich 的编译与安装，掌握 mpich 的测试与运行。

4.3.2 相关知识

1．MPI 简介

MPI（Message Passing Interface，消息传递接口）是一个消息传递库，也是一种消息传递基本模型，并成为这种编程模型的代表和事实上的标准。MPI 本身并不是一个具体的实现，而只是一种标准或规范的代表，1994 年 5 月标准的 1.0 版本诞生。MPI 是一个跨语言的通讯协议，用于编写并行计算机，支持点对点和广播。MPI 是一个信息传递应用程序接口，包括协议和和语义说明，他们指明其如何在各种实现中发挥其特性。MPI 的目标是高性能、大规模性和可移植性。MPI 在今天仍为高性能计算的主要模型。由于 MPI 是一个库而不是一门语言，因此对 MPI 的使用必须和特定的语言结合起来进行。FORTRAN 是科学与工程计算的领域语言，而 C 又是目前使用最广泛的系统和应用程序开发的语言之一，因此对 FORTRAN 和 C 的支持是必须的。

MPICH 的开发主要是由 Argonne National Laboratory 和 Mississippi State University 共同完成的，在这一过程中 IBM 也做出了自己的贡献，但是 MPI 规范的标准化工作是由 MPI 论坛完成的。MPICH 是 MPI 最流行的非专利实现，由 Argonne 国家实验室和密西西比州立大学联合开发，具有更好的可移植性，现阶段多流行的是 MPICH2。MPICH 的开发与 MPI 规范的制订是同步进行的，因此 MPICH 最能反映 MPI 的变化和发展。MPICH 是 MPI 标准的一种最重要的实现，可以免费从其官方网站下载：http://www.mpich.org/。

2．MPI 的特点

MPI 的核心工作就是实现大量服务器计算资源的整合输出，这对云计算尤为重要。MPI 的一个最重要的特点就是免费和源代码开放。它采用广为使用的语言 FORTRAN 和 C 进行绑定。MPI 为自己制定了一个雄心勃勃的目标，总结概括起来，它包括几个在实际使用中都十分重要但有时又是相互矛盾的 3 个方面：较高的通信性能；较好的程序可移植性和强大的功能。MPI 为分布式程序设计人员提供了最大的灵活性和自由度，但随之而来的代价是编程的复杂性，同时网络也是 MPI 的主要瓶颈。目前 MPI 的应用领域主要还是科学计算领域，但随着云计算与大数据技术的发展和普及，这种分布式计算机制也越来越受关注。具体地说，MPI 的目的体现在如下几方面。

- 提供应用程序编程接口。

- 提高通信效率，措施包括避免存储器到存储器的多次重复复制，允许计算和通信的重叠等。
- 可在异构环境下提供实现。
- 提供的接口可以方便 C 语言和 Fortran 语言的调用。
- 提供可靠的通信接口，即用户不必处理通信失败。
- 定义的接口和现在已有接口差别不能太大，但是允许扩展以提供更大的灵活性。
- 定义的接口能在基本的通信和系统软件无重大改变时，在许多并行计算机生产商的平台上实现，接口的语义是独立于语言的。

4.3.3 任务实施

分别在 node1、node2、node3、node4 节点机上安装开发工具。

步骤 1

安装开发包，本实训使用系统光盘安装。

（1）安装 gcc 开发包，操作命令如下。

```
[root@node1 ~]#cd /media/CentOS_6.5_Final/Packages/
[root@node1 Packages]#rpm -ivh mpfr-2.4.1-6.el6.x86_64.rpm
[root@node1 Packages]#rpm -ivh cpp-4.4.7-4.el6.x86_64.rpm
[root@node1 Packages]#rpm -ivh ppl-0.10.2-11.el6.x86_64.rpm
[root@node1 Packages]#rpm -ivh cloog-ppl-0.15.7-1.2.el6.x86_64.rpm
[root@node1 Packages]#rpm -ivh gcc-4.4.7-4.el6.x86_64.rpm
```

（2）安装 gcc-c++ 开发包，操作命令如下。

```
[root@node1 Packages]#rpm -ivh libstdc++-4.4.7-4.el6.x86_64.rpm
[root@node1 Packages]#rpm -ivh libstdc++-devel-4.4.7-4.el6.x86_64.rpm
[root@node1 Packages]#rpm -ivh gcc-c++-4.4.7-4.el6.x86_64.rpm
```

（3）安装 gcc-gfortran 开发包，操作命令如下。

```
[root@node1 Packages]#rpm -ivh libgfortran-4.4.7-4.el6.x86_64.rpm
[root@node1 Packages]#rpm -ivh gcc-gfortran-4.4.7-4.el6.x86_64.rpm
```

（4）安装 Java-1.7.0 开发包，操作命令如下。

```
[root@node1 Packages]#rpm -ivh java-1.7.0-openjdk-devel-1.7.0.45-2.4.3.3.el6.x86_64.rpm
```

Java-1.7.0 默认的工作目录为

```
/usr/lib/jvm/java-1.7.0
```

另外，如果连上因特网，则开发包可以使用 yum 安装。安装命令如下。

```
yum -y install gcc
yum -y install gcc-c++
yum -y install gcc-gfortran
yum -y install java-1.7.0-openjdk*
```

步骤 2

MPICH 下载地址为

http://www.mpich.org/static/downloads/3.1.3/mpich-3.1.3.tar.gz

如果在 Windows 系统下载，可以使用 WinSCP 或 FlashFXP 上传 mpich-3.1.3.tar.gz 软件包到 node1 节点机的 root 目录下。

WINSCP 上传文件操作如下。

（1）打开 WinSCP，输入主机名或 IP 地址和用户名密码登录，如图 4.1 所示。

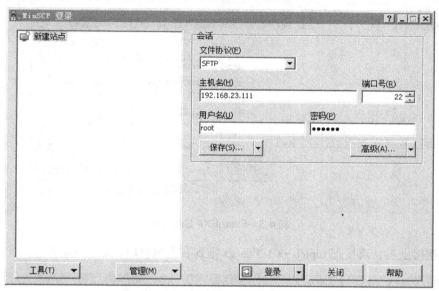

图 4.1　WinSCP 登录界面

（2）登录后进入操作界面，直接把 mpich-3.1.3.tar.gz 拖到右边窗口即可，如图 4.2 所示。

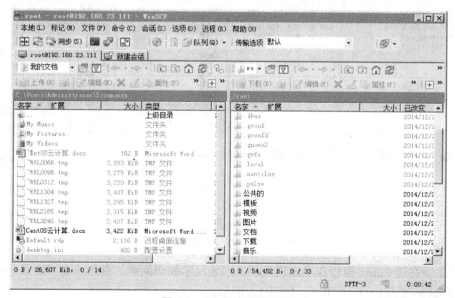

图 4.2　WinSCP 操作界面

FlashFXP 上传文件操作如下。

(1) 打开 FlashFXP，单击"快速连接"图标，选择连接类型，输入 IP 地址、用户名、密码，如图 4.3 所示。

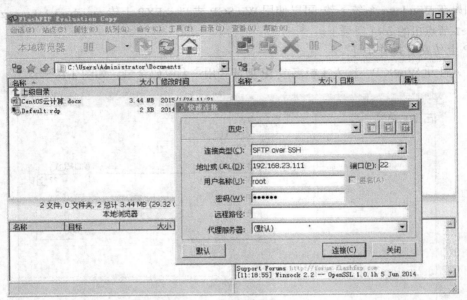

图 4.3　FlashFXP 连接界面

(2) 连接上后，直接把 mpich-3.1.3.tar.gz 拖到右上侧窗口，如图 4.4 所示。

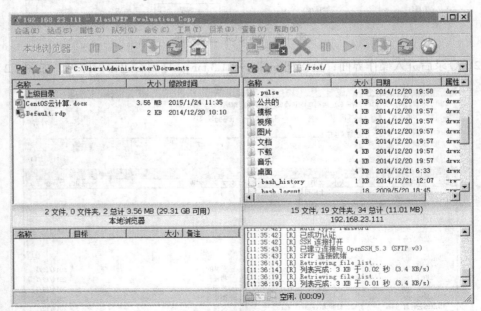

图 4.4　FlashFXP 操作界面

(3) 如果右边 Linux 窗口汉字出现乱码，则是汉字编码问题，直接单击"站点——站点管理"，字符编码改为 UTF-8 即可，如图 4.5 所示。

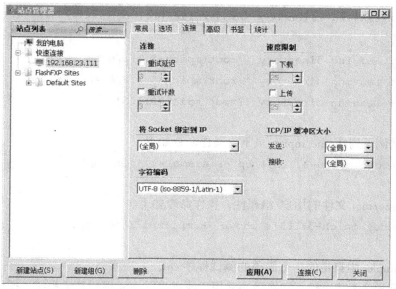

图 4.5 字符编码改为 UTF-8

步骤 3

解压 mpich 压缩包,操作命令如下。

[root@node1 ~]# tar xvzf mpich-3.1.3.tar.gz

步骤 4

进入 MPICH 解压后的目录,执行配置操作。操作命令如下。

[root@node1 ~]# cd /root/mpich-3.1.3

[root@node1 mpich-3.1.3]# ./configure --prefix=/home/mpi

配置主要的作用是对即将安装的软件进行配置,检查当前的环境是否满足要安装软件的依赖关系。参数:--prefix=PREFIX 表示把所有文件装在目录 PREFIX 下而不是默认目录下。本系统安装目录为/home/mpi。配置时间较长,请稍等,配置成功后,则最后一行提示显示"Configuration completed"。

步骤 5

执行 make 指令进行编译,编译时间较长。操作命令如下。

[root@node1 mpich-3.1.3]# make

编译后,最后显示信息:

```
  CC       cpi.o
  CCLD     cpi
make[2]: Leaving directory '/root/mpich-3.1.3/examples'
make[1]: Leaving directory '/root/mpich-3.1.3'
```

步骤 6

执行 make install 进行安装,操作命令如下。

[root@node1 mpich-3.1.3]# make install

安装后,最后显示信息:

```
make[3]: Leaving directory '/root/mpich-3.1.3'
......
make[3]: Leaving directory '/root/mpich-3.1.3/examples'
make[2]: Leaving directory '/root/mpich-3.1.3/examples'
make[1]: Leaving directory '/root/mpich-3.1.3'
```

步骤 7

复制测试例子到 /home/mpi 目录下，操作命令如下。

```
[root@node1 mpich-3.1.3]# cp -r examples /home/mpi
```

步骤 8

修改 /home/mpi 文件的用户/组属性，操作命令如下。

```
[root@node1 mpich-3.1.3]# chown -R mpi:mpi /home/mpi
```

步骤 9

运行 examples 中自带的一个简单的测试程序 cpi。

（1）登录 node1 节点机，切换到 mpi 用户或直接以 mpi 用户登录系统。操作命令如下。

```
[root@node1 mpich-3.1.3]# su - mpi
[mpi@node1~]$ _
```

（2）测试在单一节点机运行。

测试 6 个进程在单一节点机上运行：

```
[mpi@node1 ~]$ cd ~/examples/
[mpi@node1 examples]$ mpirun -np 6 ./cpi
Process 0 of 6 is on node1
Process 1 of 6 is on node1
Process 2 of 6 is on node1
Process 3 of 6 is on node1
Process 4 of 6 is on node1
Process 5 of 6 is on node1
pi is approximately 3.1415926544231239, Error is 0.0000000008333307
wall clock time = 0.000253
```

以上表示 6 个进程都在 node1 上运行。

（3）测试在 4 个节点机运行。

6 个进程在不同权重的节点机上运行，先设置 4 台节点机运行的权重值，设置如下。

```
[mpi@node1 examples]$ vi nodes
node1:2
node2:1
node3:2
node4:1
```

并行程序运行显示：

```
[mpi@node1 examples]$ mpirun -np 6 -f nodes ./cpi
```

```
Process 2 of 6 is on node2
Process 0 of 6 is on node1
Process 5 of 6 is on node4
Process 1 of 6 is on node1
Process 3 of 6 is on node3
Process 4 of 6 is on node3
pi is approximately 3.1415926544231243, Error is 0.0000000008333312
wall clock time = 0.027314
```

以上表示 6 个进程在不同节点机上运行。

（4）单进程运行程序，运行显示：

```
[mpi@node1 examples]$ ./cpi
Process 0 of 1 is on node1
pi is approximately 3.1415926544231341, Error is 0.0000000008333410
wall clock time = 0.000199
```

最后的运行结果请读者自己分析运行。

4.4 本章小结

大量的云计算系统是基于集群系统的并行计算系统。MPI 的一个最重要的特点就是免费和源代码开放。MPI 可以被迅速接受和它为自己定下的高效率、方便移植和功能强大三个主要目标密不可分。它采用广为使用的语言 FORTRAN 和 C 进行绑定。MPICH 并行环境的建立主要完成三项：配置好 NFS 服务、配置好节点间的互信、编译安装配置 MPICH。通过本章的学习，我们将理解 MPI 的概念，并掌握 MPICH 并行环境的搭建方法。

第 5 章 MPI 分布式程序设计基础

云计算是一种计算模式,代表了在某种程度上共享资源进行设计、开发、部署、运行应用,以及资源的可扩展收缩和对应用连续性的支持。云计算其实并不是一种新的技术,而是一种新的思想方法。从概念上讲,可把云计算看成是"存储云+计算云"的有机结合,即"云计算=存储云+计算云"。存储云的基础技术是分布存储,而计算云的基础技术是分布计算,更准确地说是并行计算。云计算的基础架构首先是要确保能实现并行计算。并行计算的作用是将大型的计算任务拆分,然后再派发到云中的各个节点进行分布式的并行计算,最终再将结果收集后统一处理。了解并行计算时代的程序设计方法对我们理解云计算中的一些技术基础和理念是有好处的。本章将介绍采用 MPI 进行并行程序设计的基础,使读者能领会"云计算"和"计算云"之间的联系。

5.1 任务一 最简单的并行程序的编写

5.1.1 任务描述

本节任务完成一个最简单的并行程序。这段程序虽然简单,主程序只有三行,但它确实实现了多个计算节点的共同工作,也是一个真正意义上的并行程序。通过本节的学习,读者能理解 MPI 程序的基本框架,使读者对 MPI 并行程序有一个基本的感性认识,从而了解并行程序的开发思路,掌握并行程序的开发方法,掌握并行程序的编译与运行。

5.1.2 相关知识

1. MPI 函数说明

(1)并行初始化函数:int MPI_Init(int *argc, char ***argv)。

参数描述:argc 为变量数目,argv 为变量数组,两个参数均来自 main 函数的参数。int MPI_Init 是 MPI 程序的第一个调用,它完成 MPI 程序的所有初始化工作。所有的 MPI 程序的第一条可执行语句都是这条语句。这条语句是并行代码之前第一个 mpi 函数(除 MPI_Initialize()外),启动 MPI 环境,标志并行代码的开始,要求 main 必须带参数允许,否则出错。

MPI_Init( )是 MPI 程序的第一个函数调用,标志着并行程序部分的开始,它完成 MPI 程序的初始化工作,所有 MPI 程序并行部分的第一条可执行语句都是这条语句。该函数的返回值为调用成功标志。同一个程序中 MPI_Init( )只能被调用一次。函数的参数为 main 函数的参数地址，所以并行程序和一般 C 语言程序不一样，它的 main 函数参数是不可缺少的，因为 MPI_Init( )函数会用到 main 函数的两个参数。

（2）并行结束函数：int MPI_Finalize( )。

MPI_Finalize 是 MPI 程序的最后一个调用，它结束 MPI 程序的运行，它是 MPI 程序的最后一条可执行语句，否则程序的运行结果是不可预知的。这条语句标志并行代码的结束，结束除主进程之外其他进程。一旦调用该函数后，将不能再调用其他的 MPI 函数，此时程序将释放 MPI 的数据结构及操作。这条语句之后串行代码仍可在主进程上运行(如果必须)。该函数的调用较简单，没有参数。

2. MPI 程序的编译与运行

（1）MPI 程序编译命令。

MPI 程序编译命令，常用参数：-g 加入调试信息。这条命令在联接时可以自动提供 MPI 需要的库并提供特定的开关选项。

```
mpicc /mpicc/mpif77/mpif90
```

mpicc 编译并联接用 C++编写的 MPI 程序，而 mpicc 是编译并联接用 C 编写的 MPI 程序。mpif77 和 mpif90 分别编译并联接用 FORTRAN77 和 Fortran90 编写的 MPI 程序。本书采用 mpicc 命令编译并联接用 C 编写的 MPI 程序。

常用的编译选项有：

-mpilog   产生 MPI 的 log 文件；

-mpitrace 产生跟踪文件；

-mpitrace 在该 MPI 程序执行时会打印出其运行踪迹信息，但是它和-mpilog 在编译时不能同时存在，只能二者选一；

-mpianim 产生实时动画；

-show   显示编译时产生的命令，但并不执行它；

-help   给出帮助信息；

-echo   显示出当前正在编译联接的命令信息。

此外它们还可以使用一般的 C++/C/FORTRAN77/Fortran90 通用的选项，含义和原来的编译器相同。

● 编译一个简单的 hello.c 程序。命令：

```
mpicc -c foo.c
```

● 联接输出并生成一个可执行文件。命令：

```
mpicc -o hello hello.o
```

● 上述编译联接两条语句可以采用下面一条语句。命令：

```
mpicc -o hello hello.c
```

（2）MPI 程序的运行。

最简单的 MPI 运行命令：

```
mpirun -np N program
```

其中：program 是可执行 MPI 程序名以这种方式进行执行，N 是同时运行的进程的个数。如果以一定权重来运行 MPI 程序，我们可以先写一个文件保存权重值（如 nodes），然后以一定权重值运行 MPI 程序。格式：

```
mpirun -np N -f nodes ./program
```

5.1.3 任务实施

步骤 1

编写并行源代码，操作如下。

```
[mpi@node1 examples]$ vihello.c
```

输入代码：

```c
/* 文件名：hello.c */
#include <stdio.h>
#include "mpi.h"
int main(int argc,char **argv)
{
    MPI_Init(&argc,&argv); //并行部分开始
    printf("hello paralle world!");
    printf("汕尾职业技术学院云工作室。\n");
    MPI_Finalize(); //并行结束
}
```

步骤 2

编译运行 hello.c 程序。

```
[mpi@node1 examples]$ mpicc -o hello hello.c
```

运行结果：

```
[mpi@node1 examples]$ mpirun -n 4 -f nodes ./hello
hello paralle world! 汕尾职业技术学院云工作室
hello paralle world! 汕尾职业技术学院云工作室
hello paralle world! 汕尾职业技术学院云工作室
hello paralle world! 汕尾职业技术学院云工作室
```

5.2 任务二 获取进程标志和机器名的并行程序的编写

5.2.1 任务描述

并行程序设计需要协调大量的计算节点参与计算，而且需要将任务分配到各个节点并实现节点间的数据和信息交换，面对成百上千的不同节点如没有有效的管理将面临计算的混乱，并行计算的实现将无法完成，因此各个进程需要对自己和其他进程进行识别和管理，每个进程都需要有一个唯一的 ID，用于并行程序解决"我是谁"的问题，从而实现对大量计算节点的管理和控制，有效地完成并行计算任务。因此获取进程标识和机器名是 MPI 需要完成的基

本任务，各节点根据自己的进程 ID 判断哪些任务需要自己完成。本节任务主要完成获取进程标志和机器名的并行程序的编写。

5.2.2 相关知识

（1）获得当前进程标识函数：int MPI_Comm_rank (MPI_Comm comm, int *rank)。

参数描述：comm 为该进程所在的通信域句柄；rank 为调用这一函数返回的进程在通信域中的标识号。

当 MPI 初始化后，每一个活动进程变成了一个叫作 MPI_COMM_WORLD 通信域中的成员。通信域是一个不透明对象，提供了在进程之间传递消息的环境。在一个通信域内的进程是有序的。每一个进程有一个唯一的序号（rank 号）。有了这一标识号，不同的进程就可以将自身和其他的进程区别开来，节点间的信息传递和协调均需要这一标识号。进程可以通过调用函数 MPI_Comm_rank 来确定它在通信域中的序号。

（2）获取通信域包含的进程总数函数：int MPI_Comm_size(MPI_Comm comm, int *size)。

参数描述：comm 为通信域句柄，size 为函数返回的通信域 comm 内包括的进程总数。进程通过调用 MPI_Comm_size 来确定一个通信域中的进程总数。

（3）获得本进程的机器名函数：int MPI_Get_processor_name(char *name,int *resultlen)。

参数描述：name 为返回的机器名字符串，resultlen 为返回的机器名长度。

这个函数通过字符指针*name、整型指针*resultlen 返回机器名及机器名字符串的长度。MPI_MAX_PROCESSOR_NAME 为机器名字符串的最大长度，它的值为 128。

5.2.3 任务实施

步骤 1

编辑文件 sw1.c，操作如下。

```
[mpi@node1 examples]$ visw1.c
```

输入代码：

```c
/* 文件名 sw1.c */
#include <stdio.h>
#include "mpi.h"

int main( int argc, char *argv[] )
{
    int rank;
    int size;
    char pcname[ MPI_MAX_PROCESSOR_NAME];
    int  pcnamelen;

    MPI_Init( 0, 0 );
    MPI_Comm_rank(MPI_COMM_WORLD, &rank);    //获得本进程 ID
    MPI_Comm_size(MPI_COMM_WORLD, &size);       //获得总的进程数目
```

```
    MPI_Get_processor_name(pcname,&pcnamelen); //获得本进程到机器名
    printf( "汕尾职业技术学院 from process %d of %d on %s\n", rank, size, pcname);
    MPI_Finalize();
    return 0;
}
```

步骤 2

编译 sw1.c，操作如下。

```
[mpi@node1 examples]$ mpicc -o sw1 sw1.c
```

（1）4 个进程在不同节点机上运行，运行结果：

```
[mpi@node1 examples]$ mpirun -n 4 -f nodes ./sw1
汕尾职业技术学院 from process 0 of 4 on node1
汕尾职业技术学院 from process 1 of 4 on node1
汕尾职业技术学院 from process 2 of 4 on node2
汕尾职业技术学院 from process 3 of 4 on node3
```

（2）6 个进程在不同节点机运行，运行结果：

```
[mpi@node1 examples]$ mpirun -n 6 -f nodes ./sw1
汕尾职业技术学院 from process 0 of 6 on node1
汕尾职业技术学院 from process 1 of 6 on node1
汕尾职业技术学院 from process 2 of 6 on node2
汕尾职业技术学院 from process 5 of 6 on node4
汕尾职业技术学院 from process 3 of 6 on node3
汕尾职业技术学院 from process 4 of 6 on node3
```

5.3 任务三 有消息传递功能的并行程序的编写

5.3.1 任务描述

消息传递是 MPI 编程的核心功能，也是基于 MPI 编程的设计人员需要深刻理解的功能，由于 MPI 的消息传递功能为我们提供了灵活方便的节点间数据交换和控制能力，掌握好 MPI 消息传递编程方法就掌握了 MPI 并行程序设计的核心。MPI 为程序设计者提供了丰富的消息传递函数封装。本节任务是编写简单的消息传递功能的并行程序。通过本节的学习，读者将了解并行计算的消息传递，掌握带有消息传递功能的并行程序的编写方法。

5.3.2 相关知识

1. 消息发送函数 int MPI_Send(void* buf, int count, MPI_Datatype datatype, int dest, int tag, MPI_Comm comm)

参数描述：buf 为发送缓冲区的起始地址，count 将发送的数据的个数，datatype 发送数据的数据类型；dest 为目的进程标识号；tag 为消息标志；comm 为通信域。

MPI_Send()函数是 MPI 中的一个基本消息发送函数,实现了消息的阻塞发送,在消息未发送完时程序处于阻塞状态。MPI_Send()将发送缓冲区 buf 中的 count 个 datatype 数据类型的数据发送到目的进程,目的进程在通信域中的标识号是 dest,本次发送的消息标志是 tag,使用这一标志就可以把本次发送的消息和本进程向同一目的进程发送的其他消息区别开来。MPI_Send()操作指定的发送缓冲区是由 count 个类型为 datatype 的连续数据空间组成,起始地址为 buf。注意,这里 count 的值不是以字节计数,而是以数据类型为单位指定消息的长度,这样就独立于具体的实现,并且更接近于用户的观点,发送 10 个 MPI_FLOAT 型的数据,则 count 应为 10,而不是所占的字节数。其中 datatype 数据类型可以是 MPI 的预定义类型,也可以是用户自定义的类型,但不能直接使用 C 语言中的数据类型。

部分 C 语言中的数据类型和 MPI 预定义的数据类型对比如表 5.1 所示。

表 5.1 数据类型对比

MPI 预定义数据类型	C 语言数据类型
MPI_CHAR	signed char
MPI_SHORT	signed short int
MPI_INT	signed int
MPI_LONG	signed long int
MPI_UNSIGNED_CHAR	unsigned char
MPI_UNSIGNED_SHORT	unsigned short int
MPI_UNSIGNED	unsigned int
MPI_UNSIGNED_LONG	unsigned long int
MPI_FLOAT	float
MPI_DOUBLE	double
MPI_LONG_DOUBLE	long double

2. 消息接收函数 int MPI_Recv(void* buf, int count, MPI_Datatype datatype, int source, int tag, MPI_Comm comm, MPI_Status *status)

参数描述:buf 为接收缓冲区的起始地址;count 为最多可接收的数据个数;datatype 为接收数据的数据类型;source 为接收数据的来源进程标识号;tag 为消息标识,应与相应发送操作的标识相匹配;comm 为本进程和发送进程所在的通信域;status 为返回状态。

MPI_Recv()是 MPI 中基本的消息接收函数,MPI_Recv()从指定的进程 source 接收消息,并且该消息的数据类型和消息标识与该接收进程指定的 datatype 和 tag 相一致,接收到的消息所包含的数据元素的个数最多不能超过 count。接收缓冲区是由 count 个类型为 datatype 的连续元素空间组成,由 datatype 指定其类型,起始地址为 buf,count 和 datatype 共同决定了接收缓冲区的大小,接收到的消息长度必须小于或等于接收缓冲区的长度,这是因为如果接收到的数据过大,MPI 没有截断,接收缓冲区会发生溢出错误,因此编程者要保证接收缓冲区的长度不小于发送数据的长度。如果一个短于接收缓冲区的消息到达,那么只有相应于这个消息的那些地址被修改,count 可以是零,这种情况下消息的数据部分是空的。其中 datatype 数据类型可以是 MPI 的预定义类型,也可以是用户自定义的类型,通过指定不同的数据类型调用 MPI_Recv()可以接收不同类型的数据。

消息接收函数和消息发送函数的参数基本是相互对应的，只是消息接收函数多了一个 status 参数。返回状态变量 status 用途很广，它是 MPI 定义的一个数据类型，使用之前需要用户为它分配空间。在 C 语言实现中，状态变量是由至少 3 个域组成的结构类型。这 3 个域分别是 MPI_SOURCE、MPI_TAG 和 MPI_ERROR。它还可以包括其他的附加域，这样通过对 status.MPI_SOURCE、status.MPI_TAG 和 status.MPI_ERROR 的引用就可以得到返回状态中所包含的发送数据进程的标识、发送数据使用的 tag 标识和该接收操作返回的错误代码。

5.3.3 任务实施

步骤 1

编辑文件 sw2.c，操作如下。

```
[mpi@node1 examples]$ vi sw2.c
```

输入代码：

```c
/* 文件名sw2.c */
#include <stdio.h>
#include "mpi.h"

int main( int argc, char *argv[] )
{
    int rank, numprocs, source;
    MPI_Status status;
    char message[50];

    MPI_Init(&argc, &argv);
    MPI_Comm_rank(MPI_COMM_WORLD, &rank);       //获得本进程ID
    MPI_Comm_size(MPI_COMM_WORLD, &numprocs);   //获得总的进程数目

    if (rank != 0){
        strcpy(message," Hi,汕尾职业技术学院!");
        MPI_Send(message,strlen(message)+1, MPI_CHAR, 0,49, MPI_COMM_WORLD);
    }
    else  {
        for (source = 1 ;source < numprocs; source++) {
            MPI_Recv(message,50,MPI_CHAR,source,49,MPI_COMM_WORLD,&status);
            printf("我是进程%d,我从进程%d 接收到'%s'.\n", rank, source, message);
        }
    }
}
```

```
    MPI_Finalize();
    return 0;
}
```

步骤 2

编译 sw2.c 文件，操作如下。

```
[mpi@node1 examples]$ mpicc -o sw2 sw2.c
```

运行测试，分 4 个进程在不同节点机运行 sw2 程序，运行及结果：

```
[mpi@node1 examples]$ mpirun -n 4 -f nodes ./sw2
```

我是进程 0，我从进程 1 接收到'Hi,汕尾职业技术学院!'。
我是进程 0，我从进程 2 接收到'Hi,汕尾职业技术学院!'。
我是进程 0，我从进程 3 接收到'Hi,汕尾职业技术学院!'。

5.4 本章小结

本章介绍了 MPI 分布式程序设计的基础，使读者真正步入计算的云时代。本章共 3 个任务，分别介绍了最简单的并行程序、获取进程标识和机器名的并行程序和有消息传递功能的并行程序的编写。当然，MPI 分布式程序设计的知识远远不止这些，需要我们今后继续学习。

第 6 章 Hadoop 软件的编译打包

Hadoop 是一个分布式文件系统（Hadoop Distributed File System， HDFS）。HDFS 有高容错性的特点，一般部署在低廉的硬件上，它提供高吞吐量访问应用程序的数据，适合那些有着超大数据集的应用程序。在官方网站只提供编译好的 32 位 Hadoop 软件包， 64 位 Hadoop 软件包需要重新编译。

6.1 任务一 安装编译环境

6.1.1 任务描述

在官方网站下载 Hadoop 源代码，编译 Hadoop 软件包。本节任务是了解 Linux 编译命令,了解 ant 的作用，了解 maven 的用法，掌握编译环境的搭建，掌握使用 maven 对 Hadoop 软件打包。

6.1.2 相关知识

当编译 Linux 内核及一些软件的源程序时，需要一些相关命令和软件包。

1. Make

make 命令依赖 make 文件，内容如下。

```
sw: sw.c
cc -o sw sw.c
```

该文件包含两行，一行是依赖关系，一行是执行动作。依赖关系的那一行包含了程序的名字，紧跟着一个冒号，然后是空格，最后是源文件的名字。当 mak 读入这一行的时候，会检查 foo 是否存在。如果存在，就比较 sw 和 sw.c 最后的修改时间有什么不同；如果 sw 不存在，或者比 sw.c 旧，就检查执行那一行动作。

2. Maven

Maven 是一个项目管理工具，它包含了一个项目对象模型、一组标准集合、一个项目生命周期、一个依赖管理系统和用来运行定义在生命周期阶段中插件目标的逻辑。当你使用 Maven 的时候，你用一个明确定义的项目对象模型来描述你的项目，然后 Maven 可以应用横切的逻辑，这些逻辑来自一组共享的（或者自定义的）插件。Maven 有一个生命周期，当你运行 mvn install 的时候被调用。这条命令告诉 Maven 执行一系列的有序的步骤，直到到达你指定的生命周期。遍历生命周期旅途中的一个影响就是，Maven 运行了许多默认的插件目标，这些目标完成了像编译和创建一个 JAR 文件这样的工作。此外，Maven 能够很方便地帮你管理项目报告，生成站点，管理 JAR 文件，等等。

3．Apache Ant

Apache Ant 是一个将软件编译、测试、部署等步骤联系在一起加以自动化的一个工具，大多用于 Java 环境中的软件开发。由 Apache 软件基金会所提供。Ant 是纯 Java 语言编写的，所以具有很好的跨平台性。Ant 运行时需要一个 XML 文件（构建文件）。Ant 通过调用 target 树，就可以执行各种 task。每个 task 实现了特定接口对象。

4．Protobuf

Protobuf 是 Google 的一种数据交换的格式，它独立于语言，独立于平台。Google 提供了 3 种语言的实现：java、c++ 和 python，每一种实现都包含了相应语言的编译器及库文件。由于它是一种二进制的格式，比使用 xml 进行数据交换快许多，因此可用于分布式应用之间的数据通信或者异构环境下的数据交换。作为一种效率和兼容性都很优秀的二进制数据传输格式，可以用于诸如网络传输、配置文件、数据存储等诸多领域。

6.1.3 任务实施

准备工作

选择一台安装 CentOS-6.5-x86_64 系统的计算机，以 root 用户登录。

步骤 1

安装 jdk、gcc、gcc-c++、make、cmake、openssl-devel、ncurses-devel。

（1）安装 jdk，操作命令：

```
[root@localhost~]#cd /media/CentOS_6.5_Final/Packages/
[root@localhost Packages]#rpm -ivh java-1.7.0-openjdk-devel-1.7.0.45-2.4.3.3.el6.x86_64.rpm
```

（2）安装 gcc 开发包，操作命令：

```
[root@localhost Packages]#rpm -ivh mpfr-2.4.1-6.el6.x86_64.rpm
[root@localhost Packages]#rpm -ivh cpp-4.4.7-4.el6.x86_64.rpm
[root@localhost Packages]#rpm -ivh ppl-0.10.2-11.el6.x86_64.rpm
[root@localhost Packages]#rpm -ivh cloog-ppl-0.15.7-1.2.el6.x86_64.rpm
[root@localhost Packages]#rpm -ivh gcc-4.4.7-4.el6.x86_64.rpm
```

（3）安装 gcc-c++ 开发包，操作命令：

```
[root@localhost Packages]#rpm -ivh libstdc++-devel-4.4.7-4.el6.x86_64.rpm
[root@localhost Packages]#rpm -ivh gcc-c++-4.4.7-4.el6.x86_64.rpm
```

（4）安装 gcc-gfortran 开发包，操作命令：

```
[root@localhost Packages]#rpm -ivh libgfortran-4.4.7-4.el6.x86_64.rpm
[root@localhost Packages]#rpm -ivh gcc-gfortran-4.4.7-4.el6.x86_64.rpm
```

（5）安装 camke，操作命令：

[root@localhost Packages]#rpm -ivh cmake-2.6.4-5.el6.x86_64.rpm

（6）安装 openssl-devel，操作命令：

[root@localhost Packages]# rpm -ivh keyutils-libs-devel-1.4-4.el6.x86_64.rpm

[root@localhost Packages]# rpm -ivh libcom_err-devel-1.41.12-18.el6.x86_64.rpm

[root@localhost Packages]# rpm -ivh libsepol-devel-2.0.41-4.el6.x86_64.rpm

[root@localhost Packages]# rpm -ivh libselinux-devel-2.0.94-5.3.el6_4.1.x86_64.rpm

[root@localhost Packages]# rpm -ivh krb5-devel-1.10.3-10.el6_4.6.x86_64.rpm

[root@localhost Packages]# rpm -ivh zlib-devel-1.2.3-29.el6.x86_64.rpm

[root@localhost Packages]# rpm -ivh openssl-devel-1.0.1e-15.el6.x86_64.rpm

（7）安装 ncurses-devel，操作命令：

[root@localhost Packages]# rpm -ivh ncurses-devel-5.7-3.20090208.el6.x86_64.rpm

另外，如果能连上因特网，则可以使用 yum 安装开发包，安装命令如下。

```
yum install java-1.7.0-openjdk*
yum install gcc
yum install gcc-c++
yum install make
yum install cmake
yum install openssl-devel
yum install ncurses-devel
```

步骤 2

（1）下载 maven 软件包，地址：

http://mirror.bit.edu.cn/apache/maven/maven-3/3.2.5/binaries/apache-maven-3.2.5-bin.tar.gz

（2）解压 maven 软件包，操作命令：

[root@localhost~]# cd

[root@localhost~]# tar xvzf /root/apache-maven-3.2.5-bin.tar.gz

（3）把 maven 软件移到目录/usr/local 下，操作命令：

[root@localhost~]# mv apache-maven-3.2.5 /usr/local/maven

（4）编辑环境变量，操作命令：

[root@localhost~]# vi /etc/profile

文件末尾添加内容如下：

export M2_HOME=/usr/local/maven

```
export M2=$M2_HOME/bin
export MAVEN_OPTS="-Xms256m -Xmx512m"
 export PATH="$M2:$PATH"
```

（5）使环境变量生效。

```
[root@localhost~]# source /etc/profile
```

（6）检查 maven 版本，显示如下表示成功。

```
[root@localhost ~]# mvn -version
Apache Maven 3.2.5 (12a6b3acb947671f09b81f49094c53f426d8cea1;
 2014-12-15T01:29:23+08:00)
Maven home: /usr/local/maven
Java version: 1.7.0_45, vendor: Oracle Corporation
Java home: /usr/lib/jvm/java-1.7.0-openjdk-1.7.0.45.x86_64/jre
Default locale: zh_CN, platform encoding: UTF-8
OS name: "linux", version: "2.6.32-431.el6.x86_64", arch: "amd64", family:
"unix"
```

（7）修改配置文件，操作命令：

```
[root@localhost ~]# vi /usr/local/maven/conf/settings.xml
```

在<mirrors></mirrors>内添加如下内容，其他的不需改动。

```
<mirror>
<id>nexus-osc</id>
<mirrorOf>*</mirrorOf>
<name>Nexusosc</name>
<url>http://maven.oschina.net/content/groups/public/</url>
</mirror>
```

在<profiles></profiles>内添加如下内容。

```
<profile>
<id>jdk-1.7</id>
<activation>
<jdk>1.7</jdk>
</activation>
<repositories>
<repository>
<id>nexus</id>
<name>local private nexus</name>
<url>http://maven.oschina.net/content/groups/public/</url>
<releases>
<enabled>true</enabled>
</releases>
```

```xml
<snapshots>
<enabled>false</enabled>
</snapshots>
</repository>
</repositories>
<pluginRepositories>
<pluginRepository>
<id>nexus</id>
<name>local private nexus</name>
<url>http://maven.oschina.net/content/groups/public/</url>
<releases>
<enabled>true</enabled>
</releases>
<snapshots>
<enabled>false</enabled>
</snapshots>
</pluginRepository>
</pluginRepositories>
</profile>
```

步骤3

(1) 下载 protobuf 软件包,地址:

https://github.com/google/protobuf/releases/download/v2.6.1/protobuf-2.6.1.tar.bz2

(2) 解压软件包,操作命令:

```
[root@localhost ~]# cd
[root@localhost ~]# tar xvjf /root/protobuf-2.6.1.tar.bz2
```

(3) 配置、编译、安装、加载,操作命令:

```
[root@localhost ~]# cd /root/protobuf-2.6.1
[root@localhost protobuf-2.6.1]# ./configure --prefix=/usr/local/protobuf
[root@localhost protobuf-2.6.1]# make
[root@localhost protobuf-2.6.1]# make install
[root@localhost protobuf-2.6.1]# ldconfig
```

(4) 编辑环境变量,操作命令:

```
[root@localhost protobuf-2.6.1]# vi /etc/profile
```

文件末尾添加内容:

```
export LD_LIBRARY_PATH=/usr/local/protobuf
export PATH="/usr/local/protobuf/bin:$PATH"
```

(5) 使环境变量设置生效,操作命令:

```
[root@localhost protobuf-2.6.1]# source /etc/profile
```
（6）检查 protobuf 版本，显示如下表示成功：
```
[root@localhost protobuf-2.6.1]# protoc --version
libprotoc 2.6.1
```
步骤 4
（1）下载 ant 软件包，地址：
http://mirrors.cnnic.cn/apache/ant/binaries/apache-ant-1.9.4-bin.tar.bz2
（2）解压软件包，操作命令：
```
[root@localhost ~]# cd
[root@localhost ~]# tar xvjf apache-ant-1.9.4-bin.tar.bz2
```
（3）把 ant 软件移到目录/usr/local 下，操作命令：
```
[root@localhost ~]# mv apache-ant-1.9.4 /usr/local/ant
```
（4）编辑环境变量，操作命令：
```
[root@localhost ~]# vi /etc/profile
```
在 profile 文件末尾添加内容：
```
export ANT_HOME=/usr/local/ant
export PATH="$JAVA_HOME/bin:$PATH:$HADOOP_PREFIX/bin:$PATH:$M2:$PATH:$ANT_HOME/bin"
```
（5）使环境变量设置生效，操作命令：
```
[root@localhost ~]# source /etc/profile
```
（6）检查 ant 版本，显示如下表示成功：
```
[root@localhost ~]# ant -version
Apache Ant(TM) version 1.9.4 compiled on April 29 2014
```
/etc/profile 后面几行显示如下内容则表示 ant 安装成功：
```
export M2_HOME=/usr/local/maven
export M2=$M2_HOME/bin
export MAVEN_OPTS="-Xms256m -Xmx512m"
export PATH="$M2:$PATH"
export LD_LIBRARY_PATH=/usr/local/protobuf
export PATH="/usr/local/protobuf/bin:$PATH"
export ANT_HOME=/usr/local/ant
export PATH="$JAVA_HOME/bin:$PATH:$HADOOP_PREFIX/bin:$PATH:$M2:$PATH:$ANT_HOME/bin"
```

6.2 任务二 编译 Hadoop 软件

6.2.1 任务描述

Hadoop 是一个非常优秀分布式文件系统，64 位的 Hadoop 软件包需要重新编译。本节任务是使用 maven 工具编译 Hadoop 源代码并打包。

6.2.2 相关知识

Maven 是基于项目对象模型，可以通过一小段描述信息来管理项目的构建，是报告和文档的软件项目管理工具。Maven 需要 pom.xml 文件，在这 xml 文件中添加 Maven 所依赖的 Jar 的名称，也就是添加<dependency></dependency>节点。Maven 的常用命令：

```
mvn archetype: create  创建 Maven 项目
mvn compile  编译源代码
mvn deploy  发布项目
mvn test-compile  编译测试源代码
mvn test  运行应用程序中的单元测试
mvn site  生成项目相关信息的网站
mvn clean  清除项目目录中的生成结果
mvn package  根据项目生成的 jar
mvn install  在本地 Repository 中安装 jar
mvn eclipse:eclipse  生成 eclipse 项目文件
mvnjetty:run  启动 jetty 服务
mvntomcat:run  启动 tomcat 服务
mvn clean package -Dmaven.test.skip=true:清除以前的包后重新打包，跳过测试类
```

6.2.3 任务实施

步骤 1

下载 hadoop2.6 版本的开源代码软件包，地址：

```
http://mirrors.cnnic.cn/apache/hadoop/common/stable/hadoop-2.6.0-src.tar.gz
```

步骤 2

解压软件包，操作命令：

```
[root@localhost ~]# cd
[root@localhost ~]# tar xvzf hadoop-2.6.0-src.tar.gz
```

步骤 3

进入软件目录，删除编译打包信息，操作命令如下。

```
[root@localhost ~]# cd hadoop-2.6.0-src
[root@localhost hadoop-2.6.0-src]# mvn clean
```

删除编译打包信息，显示如下。

```
……
[INFO] Apache Hadoop Tools Dist ........................... SUCCESS [  0.002 s]
[INFO] Apache Hadoop Tools ................................ SUCCESS [  0.002 s]
[INFO] Apache Hadoop Distribution ......................... SUCCESS [  0.001 s]
[INFO] ------------------------------------------------------------------------
```

```
[INFO] BUILD SUCCESS
[INFO] ------------------------------------------------------------
[INFO] Total time: 04:02 min
[INFO] Finished at: 2015-01-25T09:23:35+08:00
[INFO] Final Memory: 36M/331M
[INFO] ------------------------------------------------------------
```

步骤 4

编译打包，操作命令：

```
[root@localhost hadoop-2.6.0-src]# mvn package -Pdist,native -DskipTests -Dtar
```

编译打包显示如下信息。

```
……
[INFO] Apache Hadoop Tools Dist ........................... SUCCESS [ 11.970 s]
[INFO] Apache Hadoop Tools ................................ SUCCESS [  0.030 s]
[INFO] Apache Hadoop Distribution ......................... SUCCESS [ 40.759 s]
[INFO] ------------------------------------------------------------
[INFO] BUILD SUCCESS
[INFO] ------------------------------------------------------------
[INFO] Total time: 40:44 min
[INFO] Finished at: 2015-01-25T17:49:19+08:00
[INFO] Final Memory: 140M/422M
[INFO] ------------------------------------------------------------
```

以上信息表示 hadoop 打包成功。

步骤 5

编译打包后在 hadoop-dist/targe 目录下生成相应软件包，如下所示。

```
[root@localhost target]# cd
[root@localhost ~]# cd hadoop-2.6.0-src/hadoop-dist/target/
[root@localhost target] # ls
antrun                    hadoop-2.6.0.tar.gz           maven-archiver
dist-layout-stitching.sh  hadoop-dist-2.6.0.jar         test-dir
dist-tar-stitching.sh     hadoop-dist-2.6.0-javadoc.jar
hadoop-2.6.0              javadoc-bundle-options
```

至此，hadoop-2.6.0.tar.gz 软件包编译打包完成。

6.3 本章小结

本章主要介绍 make、ant、maven 和 protobuf 管理软件包的作用和使用方法，重点介绍 maven 的使用方法及其常用的命令，读者通过了解这些命令完成对 hadoop 源代码的编译和打包。

项目 7 Hadoop 环境的搭建与管理

Hadoop 是一个分布式文件系统（Hadoop Distributed File System，HDFS）。Hadoop 的框架最核心的设计就是 HDFS 和 MapReduce。HDFS 为海量的数据提供了存储，MapReduce 为海量的数据提供了计算。Hadoop 的使用需要搭建一个完整的分布式系统。本系统以四台节点机来构建 Hadoop 集群环境，其中一个节点机作为 NameNode（即 Master 节点，也叫主节点），另外三个节点机作为 DataNode（即 Slave 节点，也叫从节点）。

7.1 任务一 Hadoop 的安装与配置

7.1.1 任务描述

HDFS 在 Master 节点启动 dfs 和 yarn 服务时，需要自动启动 Slave 节点服务，HDFS 需要通过 ssh 访问 Slave 节点机。HDFS 需要搭建多台服务器组成分布式系统，节点机间需要无密码访问。本节任务是进行 ssh 的设置、用户的创建、hadoop 参数的设置,完成 HDFS 分布式环境的搭建。

7.1.2 相关知识

1. HDFS 文件系统——体系架构

HDFS 体系架构如图 7-1 所示。

Client 从来不会从 NameNode 读和写文件数据。Client 只是询问 NameNode 它应该和哪个 DataNodes 联系。Client 在一段限定的时间内将这些信息缓存,在后续的操作中 Client 直接和 DataNodes 交互。由于 NameNode 对于读和写的操作极少，所以极大地减小了 NameNode 的工作负荷，真正提高了 NameNode 的利用性能。NameNode 保存着 3 类元数据（MetaData）：文件名和块的名字空间、从文件到块的映射、副本位置。所有的 MetaData 都放在内存中。

2. YARN 的架构

YARN 架构如图 7.2 所示。

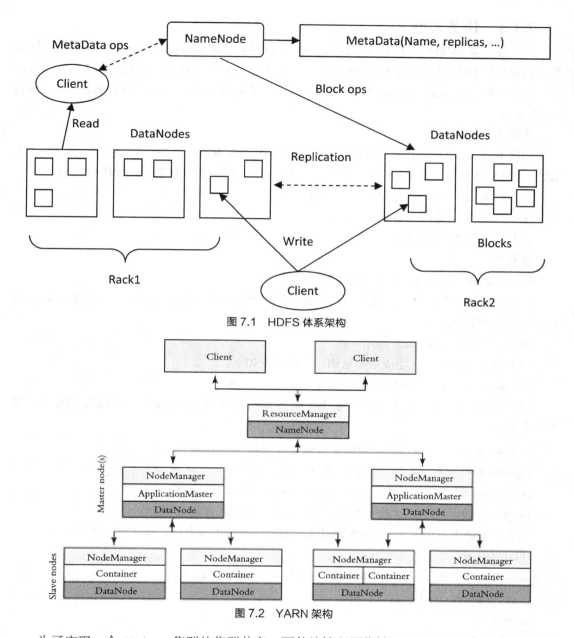

图 7.1 HDFS 体系架构

图 7.2 YARN 架构

  为了实现一个 Hadoop 集群的集群共享、可伸缩性和可靠性。YARN 采用了一种分层的集群框架方法。特定于 MapReduce 的功能已替换为一组新的守护程序，将该框架向新的处理模型开放。YARN 分层结构的本质是 ResourceManager。这个实体控制整个集群并管理应用程序向基础计算资源的分配。ResourceManager 将各个资源部分（计算、内存、带宽等）精心安排给基础 NodeManager（YARN 的每节点代理）。ResourceManager 还与 ApplicationMaster 一起分配资源，与 NodeManager 一起启动和监视它们的基础应用程序。在此上下文中，ApplicationMaster 承担了以前的 TaskTracker 的一些角色，ResourceManager 承担了 JobTracker 的角色。

7.1.3 任务实施

本节任务需要四台节点机组成集群,每个节点机上安装 CentOS-6.5-x86_64 系统。四台节点机使用的 IP 地址分别为 192.168.23.111、192.168.23.112、192.168.23.113、192.168.23.114,对应节点主机名为 node1、node2、node3、node4。节点机 node1 作为 NameNode,其他作为 DataNode。

步骤 1

创建 hadoop 用户,分别在四台节点机上创建用户 hadoop,uid=660,密码分别为 h1111、h2222、h3333、h4444。登录 node1 节点机,创建 hadoop 用户和设置密码。操作命令如下。

```
[root@node1 ~]# useradd -u 660 hadoop
[root@node1 ~]# passwd hadoop
```

其他节点机的操作相同。

步骤 2

设置 master 节点机 ssh 无密码登录 slave 节点机。

(1)在 node1 节点机上,以用户 hadoop 用户登录或者使用 su – hadoop 切换到 hadoop 用户。操作命令如下。

```
[root@node1 ~]# su - hadoop
```

(2)使用 ssh-keygen 生成证书密钥,操作命令如下。

```
[hadoop@node1 ~]$ssh-keygen -t dsa
```

(3)使用 ssh-copy-id 分别复制证书公钥到 node1,node2,node3,node4 节点机上,操作命令如下。

```
[hadoop@node1 ~]$ ssh-copy-id -i .ssh/id_dsa.pub node1
[hadoop@node1 ~]$ ssh-copy-id -i .ssh/id_dsa.pub node2
[hadoop@node1 ~]$ ssh-copy-id -i .ssh/id_dsa.pub node3
[hadoop@node1 ~]$ ssh-copy-id -i .ssh/id_dsa.pub node4
```

(4)在 node1 节点机上使用 ssh 测试无密码登录 node1 节点机,操作命令如下。

```
[hadoop@node1 ~]$ ssh node1
Last Login: Mon Dec 22 08:42:38 2014 from node1
[hadoop@node1 ~]$ exit
Logout
Connection to node1 closed.
```

以上表示操作成功。

在 node1 节点机上继续使用 ssh 测试无密码登录 node2、node3 和 node4 节点机,操作命令如下。

```
[hadoop@node1 ~]$ ssh node2
[hadoop@node1 ~]$ ssh node3
[hadoop@node1 ~]$ ssh node4
```

测试登录每个节点机后,输入 exit 退出。

步骤 3

使用 WinSCP 上传 hadoop-2.6.0.tar.gz 软件包到 node1 节点机的 root 目录下。如果 hadoop 软件包在 node1 节点机上编译，则把编译好的包复制到 root 目录下即可，如图 7.3 所示。

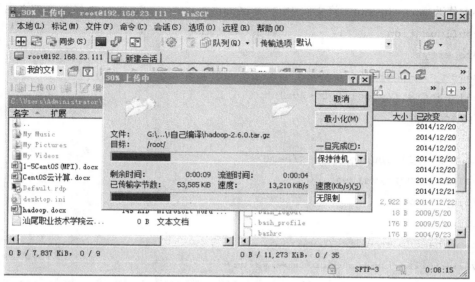

图 7.3　使用 WinSCP 上传 hadoop-2.6.0.tar.gz 软件包

步骤 4

解压文件，安装文件。操作命令如下。

```
[root@node1 ~]# cd
[root@node1 ~]# tar xvzf hadoop-2.6.0.tar.gz
[root@node1 ~]# cd hadoop-2.6.0
[root@node1 hadoop-2.6.0]# mv * /home/hadoop
```

步骤 5

修改 hadoop 配置文件。Hadoop 配置文件主要有：hadoop-env.sh、yarn-env.sh、slaves、core-site.xml、hdfs-site.xml、mapred-site.xml、yarn-site.xml。配置文件在/home/hadoop/etc/hadoop/目录下，可进入该目录进行配置。操作命令如下。

```
[root@node1 hadoop-2.6.0]# cd /home/hadoop/etc/hadoop/
```

（1）修改 hadoop-env.sh，将文件中的 export JAVA_HOME=${JAVA_HOME}

修改为 export JAVA_HOME=/usr/lib/jvm/java-1.7.0，如下所示。

```
[root@node1 hadoop]# vi hadoop-env.sh
# The java implementation to use.
# export JAVA_HOME=${JAVA_HOME}
export JAVA_HOME=/usr/lib/jvm/java-1.7.0
```

（2）修改 slaves，该文件登记 DataNode 节点主机名，本处添加 node2、node3、node4 三台节点主机名，如下所示。

```
[root@node1 hadoop]# vi slaves
```

```
node2
node3
node4
```

（3）修改 core-site.xml，将文件中的<configuration></configuration>修改为如下内容。

```
<configuration>
<property>
<name>fs.defaultFS</name>
<value>hdfs://node1:9000</value>
</property>
<property>
<name>io.file.buffer.size</name>
<value>131072</value>
</property>
<property>
<name>hadoop.tmp.dir</name>
<value>file:/home/hadoop/tmp</value>
<description>Abase for other temporary directories.</description>
</property>
<property>
<name>hadoop.proxyuser.hadoop.hosts</name>
<value>*</value>
</property>
<property>
<name>hadoop.proxyuser.hadoop.groups</name>
<value>*</value>
</property>
</configuration>
```

其中 node1 为集群的 NameNode（Master）节点机，node1 可以使用 IP 地址表示。

（4）修改 hdfs-site.xml，将文件中的<configuration></configuration>修改为如下内容。

```
<configuration>
<property>
<name>dfs.namenode.secondary.http-address</name>
<value>node1:9001</value>
</property>
<property>
<name>dfs.namenode.name.dir</name>
<value>file:/home/hadoop/dfs/name</value>
</property>
```

```xml
<property>
<name>dfs.datanode.data.dir</name>
<value>file:/home/hadoop/dfs/data</value>
</property>
<property>
<name>dfs.replication</name>
<value>3</value>
</property>
<property>
<name>dfs.webhdfs.enabled</name>
<value>true</value>
</property>
</configuration>
```

其中为了便于教学，第二个 NameNode 也使用 node1 节点机，NameNode 产生的数据存放在/home/hadoop/dfs/name 目录下，DataNode 产生的数据存放在/home/hadoop/dfs/data 目录下，设置 3 份备份。

（5）将文件 mapred-site.xml.template 改名为 mapred-site.xml，操作如下。

```
[root@node1 hadoop]# mv mapred-site.xml.template mapred-site.xml
```

将文件中的<configuration></configuration>修改为如下内容。

```xml
<configuration>
<property>
<name>mapreduce.framework.name</name>
<value>yarn</value>
</property>
<property>
<name>mapreduce.jobhistory.address</name>
<value>node1:10020</value>
</property>
<property>
<name>mapreduce.jobhistory.webapp.address</name>
<value>node1:19888</value>
</property>
</configuration>
```

（6）修改 yarn-site.xml，将文件中的<configuration></configuration>修改为如下内容。

```xml
<configuration>
<property>
<name>yarn.resourcemanager.hostname</name>
<value>192.168.23.111</value>
```

```xml
</property>
<property>
<name>yarn.nodemanager.aux-services</name>
<value>mapreduce_shuffle</value>
</property>
<property>
<name>yarn.nodemanager.aux-services.mapreduce.shuffle.class</name>
<value>org.apache.hadoop.mapred.ShuffleHandler</value>
</property>
<property>
<name>yarn.resourcemanager.address</name>
<value>node1:8032</value>
</property>
<property>
<name>yarn.resourcemanager.scheduler.address</name>
<value>node1:8030</value>
</property>
<property>
<name>yarn.resourcemanager.resource-tracker.address</name>
<value>node1:8031</value>
</property>
<property>
<name>yarn.resourcemanager.admin.address</name>
<value>node1:8033</value>
</property>
<property>
<name>yarn.resourcemanager.webapp.address</name>
<value>node1:8088</value>
</property>
</configuration>
```

步骤 6

修改"/home/hadoop/"文件用户主/组属性,操作如下。

```
[root@node1 hadoop]# chown -R hadoop:hadoop /home/hadoop
```

步骤 7

将配置好的 hadoop 系统复制到其他节点机上,操作如下。

```
[root@node1 hadoop]# cd /home/hadoop
[root@node1 hadoop]# scp -r * hadoop@node2:/home/hadoop
[root@node1 hadoop]# scp -r * hadoop@node3:/home/hadoop
```

```
[root@node1 hadoop]# scp -r * hadoop@node4:/home/hadoop
```
步骤 8

分别登录 node2、node3、node4 节点机，修改 "/home/hadoop/" 文件用户主/组属性。

```
[root@node2~]# chown -R hadoop:hadoop /home/hadoop
[root@node3~]# chown -R hadoop:hadoop /home/hadoop
[root@node4~]# chown -R hadoop:hadoop /home/hadoop
```

至此，整个 hadoop 分布式系统搭建完成。

7.2 任务二 Hadoop 的管理

7.2.1 任务描述

Hadoop 管理是不可缺少的环节，需要掌握 Hadoop 服务的启动和停止，掌握监控每台节点机的服务状态，掌握 Hadoop 应用环境的测试和使用，掌握使用浏览器监控 Hadoop 服务状态。本节任务是完成 Hadoop 的管理。

7.2.2 相关知识

1. HDFS 的存储过程

HDFS 在对一个文件进行存储时有两个重要的策略：一个是副本策略，另一个是分块策略。副本策略保证了文件存储的高可靠性，分块策略保证数据并发读写的效率并且是 MapReduce 实现并行数据处理的基础。

HDFS 的分块策略：通常 HDFS 在存储一个文件会将文件切为 64MB 大小的块来进行存储，数据块会被分别存储在不同的 Datanode 节点上。这一过程其实就是一种数据任务的切分过程，在后面对数据进行 MapReduce 操作时十分重要，同时数据被分块存储后能实现对数据的并发读写，提高数据读写效率。

2. jps 命令

jps（Java Virtual Machine Process Status Tool）是 JDK 提供的一个显示当前所有 java 进程 pid 的命令。语法格式为

```
jps [ options ] [ hostid ]
```

其中：options 可以用 -q（安静）、-m（输出传递给 main 方法的参数）、-l（显示完整路径）、-v（显示传递给 JVM 的命令行参数）、-V（显示通过 flag 文件传递给 JVM 的参数）、-J（和其他 Java 工具类似用于传递参数给命令本身要调用的 java 进程）；hostid 是主机 id，默认 localhost。

7.2.3 任务实施

步骤 1

格式化 NameNode。登录 node1 节点机，以用户 hadoop 登录或 su – hadoop 登录，格式化 NameNode。操作命令如下。

```
[root@node1~]# su - hadoop
```

```
[hadoop@node1~]$hadoop namenode -format
```
格式后，最后部分内容显示如下。

```
14/12/25 23:37:06 INFO common.Storage: Storage directory /home/hadoop/dfs/name has been successfully formatted.
14/12/25 23:37:06 INFO namenode.NNStorageRetentionManager: Going to retain 1 images with txid >= 0
14/12/25 23:37:06 INFO util.ExitUtil: Exiting with status 0
14/12/25 23:37:06 INFO namenode.NameNode: SHUTDOW_MSG:
/************************************************************
SHUTDOWN_MSG: Shutting down NameNode at node1/192.168.23.111
************************************************************/
```

最后显示有"successfully formatted."表示格式化成功。

步骤2

启动、停止 hadoop 服务。进入/home/hadoop/sbin/，可以看到目录的脚本程序，如下所示。

```
[hadoop@node1 ~]$ cd
[hadoop@node1 ~]$ cd sbin
[hadoop@node1 sbin]$ dir
distribute-exclude.sh    start-all.cmd          stop-balancer.sh
hadoop-daemon.sh         start-all.sh           stop-dfs.cmd
hadoop-daemons.sh        start-balancer.sh      stop-dfs.sh
hdfs-config.cmd          start-dfs.cmd          stop-secure-dns.sh
hdfs-config.sh           start-dfs.sh           stop-yarn.cmd
httpfs.sh                start-secure-dns.sh    stop-yarn.sh
kms.sh                   start-yarn.cmd         yarn-daemon.sh
mr-jobhistory-daemon.sh  start-yarn.sh          yarn-daemons.sh
refresh-namenodes.sh     stop-all.cmd
slaves.sh                stop-all.sh
```

（1）运行 start-dfs.sh 脚本程序，如下所示。

```
[hadoop@node1 sbin]$ ./start-dfs.sh
Starting namenodes on [node1]
  node1: starting namenode, logging to /home/hadoop/logs/hadoop-hadoop-namenode-node1.out
  node4: starting datanode, logging to /home/hadoop/logs/hadoop-hadoop-datanode-node4.out
  node3: starting datanode, logging to /home/hadoop/logs/hadoop-hadoop-datanode-node3.out
  node2: starting datanode, logging to /home/hadoop/logs/hadoop-hadoop-datanode-node2.out
```

```
Starting secondary namenodes [node1]
node1: starting secondarynamenode,
logging to /home/hadoop/logs/hadoop-hadoop-secondarynamenode-node1.out
```

（2）运行 yarn 脚本程序，如下所示。

```
[hadoop@node1 sbin]$ ./start-yarn.sh
starting yarn daemons
starting resourcemanager, logging to /home/hadoop/logs/yarn-hadoop-resourcemanager-node1.out
node2: starting nodemanager, logging to /home/hadoop/logs/yarn-hadoop-nodemanager-node2.out
node3: starting nodemanager, logging to /home/hadoop/logs/yarn-hadoop-nodemanager-node3.out
node4: starting nodemanager, logging to /home/hadoop/logs/yarn-hadoop-nodemanager-node4.out
```

（3）分别检查每台节点机运行情况，如下所示。

```
[hadoop@node1 sbin]$ jps[hadoop@node2 ~]$ jps
36059 ResourceManager38030 NodeManager
35834 SecondaryNameNode38308 Jps
35633 NameNode37855 DataNode
36738 Jps

[hadoop@node3 ~]$ jps[hadoop@node4~]$ jps
37245 DataNode37875 Jps
37417 NodeManager 37596 NodeManager
37715 Jps37424 DataNode
```

（4）停止 hadoop 服务，停止服务后，后面操作无法进行，这步暂时不操作，如下所示。

```
[hadoop@node1 sbin]$ ./stop-all.sh
This script is Deprecated. Instead use stop-dfs.sh and stop-yarn.sh
Stopping namenodes on [node1]
node1: stopping namenode
node3: stopping datanode
node2: stopping datanode
node4: stopping datanode
Stopping secondary namenodes [node1]
node1: stopping secondarynamenode
stopping yarn daemons
stopping resourcemanager
```

步骤3

查看集群状态，如下所示。

```
[hadoop@node1 ~]$ hdfs dfsadmin -report
Configured Capacity: 68788236288 (64.06 GB)
Present Capacity: 64061755392 (59.66 GB)
DFS Used: 73728 (72 KB)
DFS Used%: 0.00%
Under replicated blocks: 0
Blocks with corrupt replicas: 0
Missing blocks:0
```

步骤4

查看文件块组成，如下所示。

```
[hadoop@node1 ~]$ hdfs fsck / -files -blocks
Connecting to namenode via http://node1:50070
FSCK started by hadoop (auth:SIMPLE) from /192.168.23.111 for path / at Tue Dec 23 10:04:53 CST 2014
/ <dir>
Status: HEALTHY
 Total size:    0 B
 Total dirs:    1
 Total files:   0
 Total symlinks:        0
 Total blocks (validated):      0
 Minimally replicated blocks:   0
 Over-replicated blocks:        0
 Under-replicated blocks:       0
 Mis-replicated blocks:         0
 Default replication factor:    3
 Average block replication:     0.0
 Corrupt blocks:                0
 Missing replicas:              0
 Number of data-nodes:          3
 Number of racks:               1
FSCK ended at Tue Dec 23 10:04:53 CST 2014 in 6 milliseconds


This filesystem under path '/' is HEALTHY
```

步骤5

使用浏览器浏览 Master 节点机 http://192.168.23.111:50070，查看 NameNode 节点状态，如图 7.4 所示。

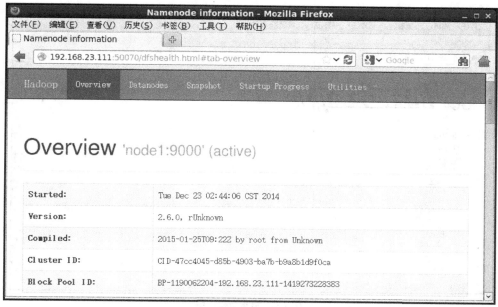

图 7.4 查看 NameNode 节点状态

步骤 6

浏览 Datanodes 数据节点，如图 7.5 所示。

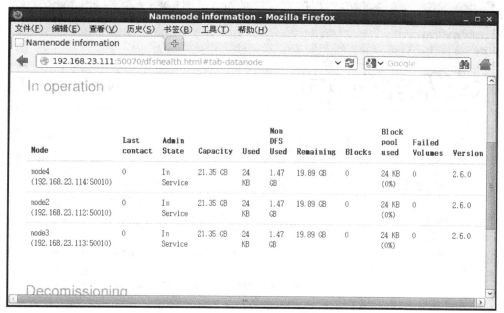

图 7.5 浏览 Datanodes 数据节点

步骤 7

使用浏览器浏览 Master 节点机 http://192.168.23.111:8088 查看所有应用，如图 7.6 所示。

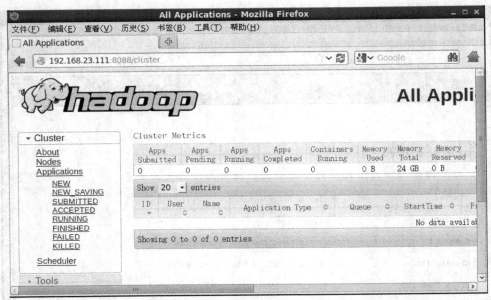

图 7.6　查看所有应用

步骤 8

浏览 Nodes，如图 7.7 所示。

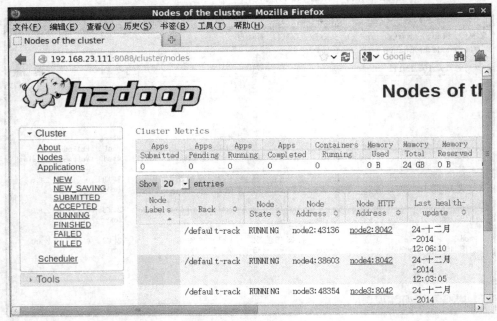

图 7.7　浏览 Nodes

7.3　任务三　Hadoop Shell 命令的使用

7.3.1　任务描述

Hadoop 提供与 hadoop 分布式文件系统交互的命令，通过了解 Hadoop Shell 命令的用法，掌握对 hadoop 分布式文件系统的操作。本节任务是使用 Hadoop Shell 命令进行操作。

7.3.2 相关知识

1. 分布式文件系统

分布式文件系统（Distributed File System）是指文件系统管理的物理存储资源不一定直接连接在本地节点上，而是通过计算机网络与节点相连搜索。对超大规模数据集提供可靠的存储功能，并对用户应用程序提供高带宽的输入输出数据流。在大型的集群里，上千台服务器均可直接参与到数据存储和应用程序任务执行。通过多服务器，分布式的存储和计算，计算资源的规模能够按照需要增长，并兼顾在各种规模上的经济适用性。

2. Hadoop Shell

为方便对 hdfs 文件系统和作业提交的操作，hadoop 提供了一些基本的 shell 操作，这些基本操作与 Linux 下的操作有很多相似性。shell 操作的基本命令格式为

```
hdfs dfs -cmd <args>
```

其中，-cmd 为具体的操作命令，如-cp（复制命令）。

7.3.3 任务实施

步骤 1

在 hdfs 创建 swvtc 目录，操作命令如下。

```
[hadoop@node1 ~]$ hdfs dfs -mkdir /swvtc        #类似 mkdir /swvtc
```

步骤 2

在 hdfs 查看当前目录，操作命令如下。

```
[hadoop@node1 ~]$ hdfs dfs -ls /                #类似 ls /
Found 1 items
drwxr-xr-x   - hadoop supergroup          0 2014-12-23 10:07 /swvtc
```

步骤 3

在本地系统编辑文件 jie.txt，操作命令如下。

```
[hadoop@node1 ~]$ vi jie.txt
```

添加内容：

Hi，汕尾职业技术学院欢迎您！

上传文件 jie.txt 到 hdfs 的/swvtc 目录中，操作命令如下。

```
[hadoop@node1 ~]$ hdfs dfs -put jie.txt /swvtc
```

步骤 4

从 hdfs 中下载文件，操作命令：

```
[hadoop@node1 ~]$ hdfs dfs -get /swvtc/jie.txt
```

步骤 5

查看 hdfs 中/swvtc/jie.txt 的内容，操作命令：

```
[hadoop@node1 ~]$ hdfs dfs -text /swvtc/jie.txt
Hi，汕尾职业技术学院欢迎您！
```

步骤 6

选 Hadoop 常用的一些命令做练习，介绍如下。

```
hdfs dfs -ls <path>                    #查看指定目录下文件
hdfs dfs -cat <path-file>              #查看文件内容
hdfs dfs -put <local-file/path><hdoop-path>  #将本地文件或目录存储至 hadoop
hdfs dfs -get <hadoop-path><local>     #将 hadoop 文件下载至本地
hdfs dfs -rm <hadoop-file>             #删除 hadoop 文件
hdfs dfs -rmr <hadoop-path>            #删除 hadoop 上指定文件夹（包含子目录等）
hdfs dfs -mkdir <hadoop-path>          #在 hadoop 创建新目录
hdfs dfs -touchz <hadoop-path-file>    #在 hadoop 新建一个空文件
hdfs dfs -mv <old-path-file><new-path-file>  #将 hadoop 文件重命名
hdfs dfs -getmerge <hadoop-path><local-path-file>  #将 hadoop 指定目录下所有内容
                                       #保存为一个文件，同时下载至本地
hadoop job -kill <job-id>              #将正在运行的 hadoop 作业杀掉
hadoop jar <jar><mainClass] args       #运行 jar 文件
hadoop distcp <srcurl><desturl>        #递归地复制文件或目录
```

步骤 7

选 Hadoop 其他一些命令做练习，介绍如下。

（1）运行 HDFS 文件系统检查工具(fsck tools)。

用法：hadoop fsck [GENERIC_OPTIONS] <path> [-move | -delete | -openforwrite] [-files [-blocks [-locations | -racks]]]

选项：

<path>检查的起始目录。

-move 移动受损文件到/lost+found。

-delete 删除受损文件。

-openforwrite 打印写打开的文件。

-files 打印正被检查的文件。

-blocks 打印块信息。

-locations 打印出每个块的位置信息。

-racks 打印出 data-node 的网络拓扑结构。

（2）用于和 Map Reduce 作业交互的命令(jar)。

用法：hadoop job [GENERIC_OPTIONS] [-submit <job-file>] | [-status <job-id>] | [-counter <job-id><group-name><counter-name>] | [-kill <job-id>] | [-events <job-id><from-event-#><#-of-events>] | [-history [all] <jobOutputDir>] | [-list [all]] | [-kill-task <task-id>] | [-fail-task <task-id>]

选项：

-submit <job-file>提交作业。

-status <job-id>打印 map 和 reduce 完成百分比和所有计数器。

-counter <job-id><group-name><counter-name> 打印计数器的值。
　　-kill <job-id>杀死指定作业。
　　-events <job-id><from-event-#><#-of-events> 打印给定范围内jobtracker接收到的事件细节。
　　-history [all] <jobOutputDir> -history <jobOutputDir> 打印作业的细节、失败及被杀死原因的细节。更多的关于一个作业的细节比如成功的任务，做过的任务尝试等信息可以通过指定[all]选项查看。
　　-list [all] -list all 显示所有作业。-list 只显示将要完成的作业。
　　-kill-task <task-id>杀死任务。被杀死的任务不会不利于失败尝试。
　　-fail-task <task-id>使任务失败。被失败的任务会对失败尝试不利。
　　（3）运行 pipes 作业。
　　用法：hadoop pipes [-conf<path>] [-jobconf<key=value>, <key=value>, ...] [-input <path>] [-output <path>] [-jar <jar file>] [-inputformat<class>] [-map <class>] [-partitioner<class>] [-reduce <class>] [-writer <class>] [-program <executable>] [-reduces <num>]。
　　选项：
　　-conf<path>作业的配置
　　-jobconf<key=value>, <key=value>, ... 增加/覆盖作业的配置项
　　-input <path>输入目录
　　-output <path>输出目录
　　-jar <jar file>Jar 文件名
　　-inputformat<class>InputFormat 类
　　-map <class>Java Map 类
　　-partitioner<class>Java Partitioner
　　-reduce <class>Java Reduce 类
　　-writer <class>Java RecordWriter
　　-program <executable>可执行程序的 URI
　　-reduces <num>reduce 个数
　　（4）运行一个 HDFS 的 dfsadmin 客户端。
　　用法：hadoop dfsadmin [GENERIC_OPTIONS] [-report] [-safemode enter | leave | get | wait] [-refreshNodes] [-finalizeUpgrade] [-upgradeProgress status | details | force] [-metasave filename] [-setQuota<quota><dirname>...<dirname>] [-clrQuota<dirname>...<dirname>] [-help [cmd]]
　　选项：
　　-report 报告文件系统的基本信息和统计信息。
　　-safemode enter | leave | get | wait 安全模式维护命令。
　　-refreshNodes 重新读取 hosts 和 exclude 文件，更新允许连到 Namenode 的或那些需要退出或入编的 Datanode 的集合。
　　-finalizeUpgrade 终结 HDFS 的升级操作。

−upgradeProgress status | details | force 请求当前系统的升级状态，状态的细节，或者强制升级操作进行。

−metasave filename 保存 Namenode 的主要数据结构到 hadoop.log.dir 属性指定的目录下的 <filename> 文件。

−setQuota<quota><dirname>...<dirname> 为每个目录 <dirname>设定配额<quota>。目录配额是一个长整型整数，强制限定了目录树下的名字个数。

−clrQuota<dirname>...<dirname> 为每一个目录<dirname>清除配额设定。

−help [cmd] 显示给定命令的帮助信息。

（5）运行 namenode。

用法：hadoop namenode [−format] | [−upgrade] | [−rollback] | [−finalize] | [−importCheckpoint]

选项：

−format 格式化 namenode。

−upgrade 分发新版本的 hadoop 后，namenode 应以 upgrade 选项启动。

−rollback 将 namenode 回滚到前一版本。

−finalize finalize 会删除文件系统的前一状态。

−importCheckpoint 从检查点目录装载镜像并保存到当前检查点目录。检查点目录由 fs.checkpoint.dir 指定。

7.4 本章小结

本章共 3 个任务，第一个任务完成 hadoop 分布式环境的搭建，第二个任务掌握 hadoop 分布式系统的管理和 hadoop 名称节点的格式化，第三个任务主要熟悉 hadoop shell 的使用。通过本章的学习，掌握 hadoop 分布式系统管理与使用。

第 8 章
Map/Reduce 实例

8.1 任务一 实现 Map/Reduce 的 C 语言实例

8.1.1 任务描述

Map/Reduce 操作代表了一大类的数据处理操作方式，为了让大家对 Map/Reduce 的工作过程有一个直观的了解，本节任务采用 C 语言实现了一个简单经典的 Map/Reduce 计算，计算从控制台输入的字符串中单词的个数。本节任务是完成 my_map() 和 my_reduce() 函数的编写，实现了对字符串的 Map 和 Reduce 操作。

8.1.2 相关知识

1．MapReduce 工作流程

MapReduce 是一种编程模型，用于大规模数据集（大于 1TB）的并行运算。概念 Map（映射）和 Reduce（归约）的主要思想，都是从函数式编程语言里借来的，还有从矢量编程语言里借来的特性。它极大地方便了编程人员在不会分布式并行编程的情况下，将自己的程序运行在分布式系统上。当前的软件实现是指定一个 Map（映射）函数，用来把一组键值对映射成一组新的键值对，指定并发的 Reduce（归约）函数，用来保证所有映射的键值对中的每一个共享相同的键组，MapReduce 工作流程如图 8.1 所示。

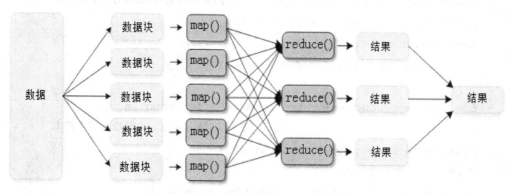

图 8.1 MapReduce 工作流程

2. MapReduce 的思想

MapReduce 的思想是"分而治之":Mapper 函数负责"分",即把复杂的任务分解为若干个"简单的任务"执行;Reducer 函数对 map 阶段的结果进行汇总。

8.1.3 任务实施

步骤 1

编写代码,如下所示。

```
[root@node1 ~]# vi mapreduce.c
```

输入代码如下。

```c
/*文件名:mapreduce.c*/
#include <stdio.h>
#include <string.h>
#include <stdlib.h>
#define BUF_SIZE    2048
int my_map(char *buffer,char (*mapbuffer)[100]);
int my_reduce(char (*mapbuffer)[100],char (*reducebuffer)[100],int *count, int num);


int main(int argc, char *argv[]){
    char buffer[BUF_SIZE];                      //定义存储字符串的缓冲区
    char mapbuffer[BUF_SIZE][100];              //定义存储 map 结果的缓冲区
    char reducebuffer[BUF_SIZE][100];           //定义存储 reduce 结果的缓冲区
    int count[BUF_SIZE]={0};                    //定义每个单词计数数组
    int num;                                    //单词总数
    int i;
    int countnum;                               //归约后的结果数
    fgets(buffer, BUF_SIZE - 1, stdin);
    buffer[strlen(buffer)-1]='\0';              //将字符串最后的回车符改为结束符
    num=my_map(buffer,mapbuffer);               //调用 map 函数处理字符串
    printf("This is map results:\n");
    for(i=0;i<num;i++) {
        printf("<%s\t1>\n",mapbuffer[i]);
    }
    countnum=my_reduce(mapbuffer,reducebuffer,count,num);
                                                //调用 reduce 函数处理字符串
    printf("This is reduce results:\n");
    for(i=0;i<countnum;i++) {
        printf("<%s\t%d>\n",reducebuffer[i],count[i]);
```

```c
    }
}
// map 函数，输入参数为字符串指针 buffer，map 后的结果通过 mapbuffer 参数传出
// 函数返回值为字符串中单词个数
int my_map(char *buffer,char (*mapbuffer)[100]){
    char *p;
    int num=0;
    if(p=strtok(buffer," ")){
        strcpy(mapbuffer[num],p);
        num++;
    }
    else
        return num;
    while(p=strtok(NULL," ")){
        strcpy(mapbuffer[num],p);
        num++;
    }
    return num;
}

// reduce 函数，输入参数为字符串 map 后的结果 mapbuffer 和单词个数 num
// reduce 结果通过 reducebuffer 和 count 参数传出
// 函数返回值为 reduce 的结果个数
int my_reduce(char (*mapbuffer)[100],char (*reducebuffer)[100],int *count,
int num){
    int i,j;
    int flag[BUF_SIZE]={0};
    char tmp[100];
    int countnum=0;
    for(i=0;i<num;i++){
        if(flag[i]==0){
            strcpy(tmp,mapbuffer[i]);
            flag[i]=1;
            strcpy(reducebuffer[countnum],mapbuffer[i]);
            count[countnum]=1;
            for(j=0;j<num;j++){
if(memcmp(tmp,mapbuffer[j],strlen(tmp))==0&&(strlen(tmp)==strlen(mapbuff
```

```
er[j]))&&(flag[j]==0))              {
                    count[countnum]++;
                    flag[j]=1;
                }
            }
        countnum++;
        }
    }
    return countnum;
}
```

步骤2

编译代码,操作如下所示。

```
[root@node1 ~]#cc -o mapreduce mapreduce.c
```

步骤3

运行程序,输入字符串。如下所示。

```
[root@node1 ~]# ./mapreduce
This is map reduce swvtc map swvtc shanwei shanwei reduce
```

运行结果如下。

```
This is map results:
<This    1>
<is      1>
<map     1>
<reduce  1>
<swvtc   1>
<map     1>
<swvtc   1>
<shanwei    1>
<shanwei    1>
<reduce  1>
This is reduce results:
<This    1>
<is      1>
<map     2>
<reduce  2>
<swvtc   2>
<shanwei    2>
```

8.2 任务二 安装 Eclipse 开发工具

8.2.1 任务描述

Eclipse 开发工具为开发 hadoop 应用程序提供便利，开发 hadoop 应用程序需要安装 hadoop 插件。本节任务是完成 hadoop 插件的安装。

8.2.2 相关知识

Eclipse 是一个开放源代码的、基于 Java 的可扩展开发平台。就其本身而言，它只是一个框架和一组服务，用于通过插件组件构建开发环境。使用 Eclipse 工具便于我们开发 hadoop 应用软件。

8.2.3 任务实施

步骤 1

（1）下载 Eclipse 软件包，地址：

http://mirrors.yun-idc.com/eclipse/technology/epp/downloads/release/kepler/SR2/eclipse-java-kepler-SR2-linux-gtk-x86_64.tar.gz

（2）解压到/usr/local 目录下，操作命令：

```
[root@node1 ~]# cd /usr/local
[root@node1 local]# tar xzvf /root/eclipse-java-kepler-SR2-linux-gtk-x86_64.tar.gz
```

步骤 2

（1）安装 hadoop 插件，下载地址：

https://github.com/winghc/hadoop2x-eclipse-plugin/archive/master.zip

（2）解压出 hadoop-eclipse-plugin-2.6.0.jar 后，放在 Eclipse 下的 plugin 目录下。操作命令：

```
[root@node1 ~]# cd /usr/local/eclipse/plugins/
[root@node1 plugins]# mv /root/hadoop-eclipse-plugin-2.6.0.jar .
```

步骤 3

运行 Eclipse，操作命令如下。

```
[root@node1 ~]# /usr/local/eclipse/eclipse &
```

执行命令后打开对话框，输入工作目录，如图 8.2 所示。

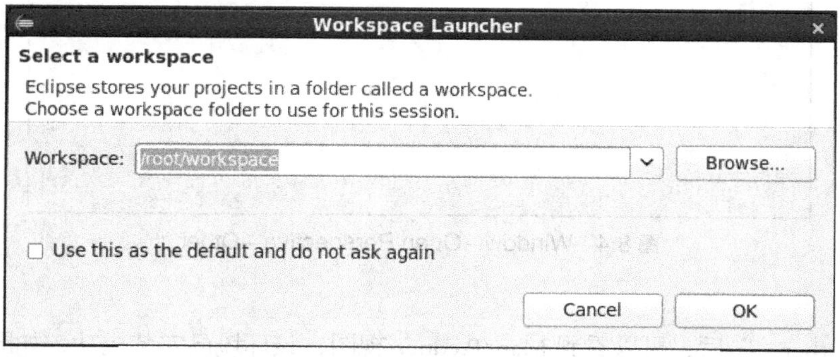

图 8.2 输入工作目录

步骤 4

单击"OK"后,进入工作界面,如图 8.3 所示。

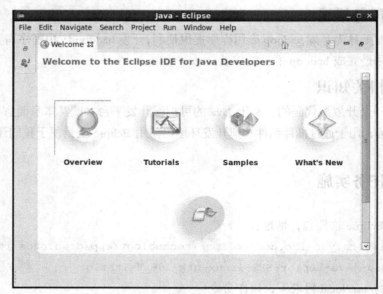

图 8.3　Eclipse 工作窗口

步骤 5

单击菜单栏的"Window—Open Perspective—Other…"菜单项,显示"Open perspective"窗口,如图 8.4 所示。

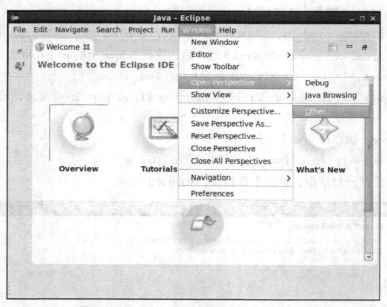

图 8.4　Window—Open Perspective—Other

步骤 6

单击"Other…"后,可以看到 Map/Reduce 视图,该环境已安装成功,如图 8.5 所示。

图 8.5 Open perspective 窗口

8.3 任务三 完成 Map/Reduce 项目

8.3.1 任务描述

实例统计 hdfs 系统文件中出现的单词数目是一个典型的 Map/Reduce 项目。本节任务是使用 Eclipse 开发工具编写一个 WordCount 项目，完成 Map/Reduce 项目的编译打包并提交运行。

8.3.2 相关知识

1．WordCount 工作流程

WordCount 顾名思义就是单词计数，用于统计文本中单词出现的次数，它是 hadoop 的入门程序。WordCount 工作流程如图 8.6 所示。

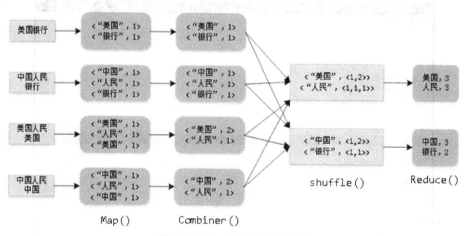

图 8.6 WordCount 工作流程

2．Eclipse 插件

Eclipse 提供了一个可扩展插件的开发系统，这就使得 Eclipse 在运行系统之上可以实现各种功能。这些插件不同于其他的应用（插件的功能是最难用代码实现的）。拥有合适的 Eclipse

插件是非常重要的，因为它们能让 Java 开发者们无缝地开发基于 J2EE 和服务的应用程序。Eclipse 的插件也能帮助他们开发不同应用架构上的程序。hadoop eclipse 插件能帮助我们完成 Map/Reduce 项目的开发。

8.3.3 任务实施

步骤 1

运行 eclipse，选择菜单栏的"File—New—Other…"菜单项，如图 8.7 所示。

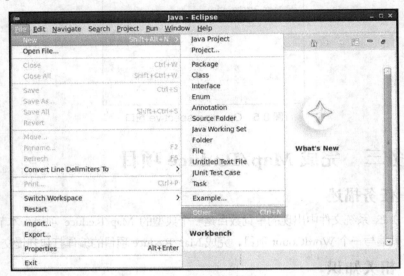

图 8.7 eclipse 操作

步骤 2

选择"Map/Reduce Project"，如图 8.8 所示。

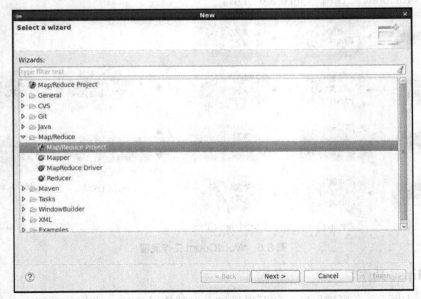

图 8.8 选择向导

步骤 3

单击"Next"按钮后进入对话框,单击"Configure Hadoop install directory…",打开对话框,输入 Hadoop 安装路径,如图 8.9 所示。

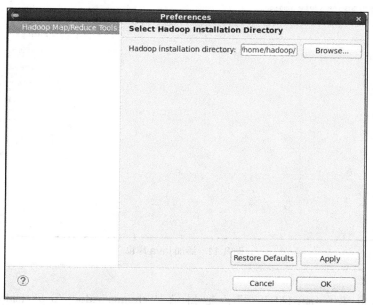

图 8.9 输入 Hadoop 安装路径

步骤 4

单击"OK"按钮,打开对话框输入项目名称 WordCount,如图 8.10 所示。

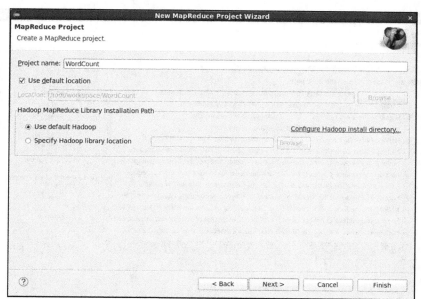

图 8.10 MapReduce 项目对话框

步骤 5

单击"Next>"按钮,选择"Libraries",如图 8.11 所示。

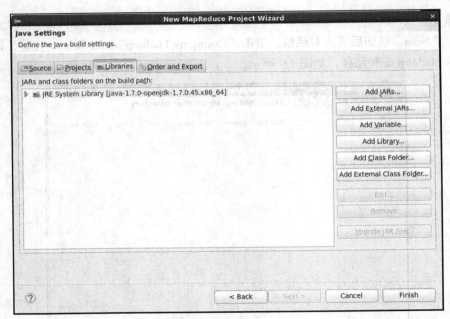

图 8.11 添加 java 库包

步骤 6

单击"Add External JARs…",把目录"/home/hadoop/share/hadoop/common"、"/home/hadoop/share/hadoop/mapreduce"、"/home/hadoop/share/hadoop/hdfs"、"/home/hadoop/share/hadoop/common/lib/commons-cli"和"/home/hadoop/share/hadoop/yarn"下的 jar 包添加到库中,如图 8.12 所示。

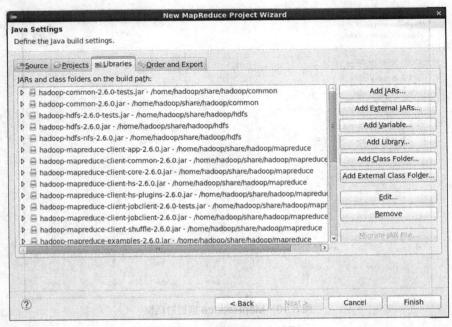

图 8.12 Add External JARs

步骤 7

单击"Finish"按钮,完成 WordCount 项目的创建,如图 8.13 所示。

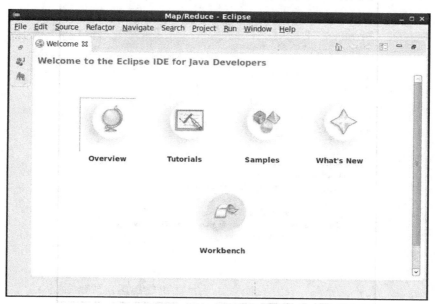

图 8.13 Map/Reduce 项目

步骤 8

单击左边大象图标,在 eclipse 的项目管理器中显示 WordCount 项目,新建 MapReduce 项目 WordCount 成功,如图 8.14 所示。

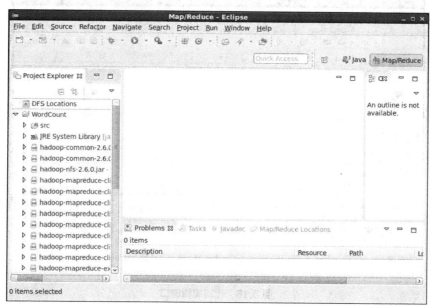

图 8.14 WordCount 项目

步骤 9

右击列表中的项目 WordCount,新建 classs,输入类名 WordCount,如图 8.15 所示。

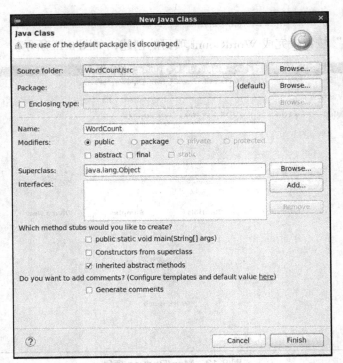

图 8.15 新建 WordCount 类

步骤 10

输入类名后,打开代码编辑窗口,如图 8.16 所示。

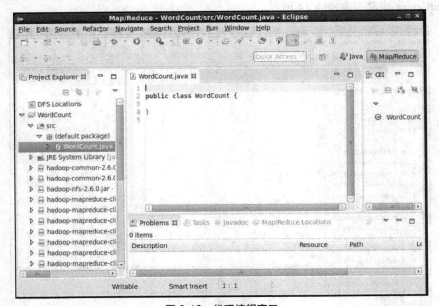

图 8.16 代码编辑窗口

输入代码,如下所示。

```
import java.io.IOException;
import java.util.StringTokenizer;
```

```java
import org.apache.hadoop.conf.Configuration;
import org.apache.hadoop.fs.Path;
import org.apache.hadoop.io.IntWritable;
import org.apache.hadoop.io.Text;
import org.apache.hadoop.mapreduce.Job;
import org.apache.hadoop.mapreduce.Mapper;
import org.apache.hadoop.mapreduce.Reducer;
import org.apache.hadoop.mapreduce.lib.input.FileInputFormat;
import org.apache.hadoop.mapreduce.lib.output.FileOutputFormat;
import org.apache.hadoop.util.GenericOptionsParser;

public class WordCount {
    //继承mapper接口，设置map的输入类型为<Object,Text>
    //输出类型为<Text,IntWritable>
    public static class Map extends Mapper<Object,Text,Text,IntWritable>{
    //one表示单词出现一次
    private static IntWritable one = new IntWritable(1);
    //word存储切下的单词
    private Text word = new Text();
        public void map(Object key,Text value,Context context) throws IOException,InterruptedException{
            //对输入的行切词
            StringTokenizer st = new StringTokenizer(value.toString());//以默认的分隔符将value切分
            while(st.hasMoreTokens()){
                word.set(st.nextToken());//切下的单词存入word
                context.write(word, one);  //将word为key，one为value写入磁盘
            }
        }
    }
    //继承reducer接口，设置reduce的输入类型<Text,IntWritable>
    //输出类型为<Text,IntWritable>
    public static class Reduce extends Reducer<Text,IntWritable,Text,IntWritable>{
        //result记录单词的频数
        private static IntWritable result = new IntWritable();
        public void reduce(Text key,Iterable<IntWritable> values,Context context) throws IOException,InterruptedException{
```

```java
        int sum = 0;
        //对获取的<key,value-list>计算value的和
        for(IntWritable val:values){
            sum += val.get();
        }
        //将频数设置到result
        result.set(sum);
        //收集结果
        context.write(key, result);
    }
}
/**
 * @param args
 */
public static void main(String[] args) throws Exception{
    // TODO Auto-generated method stub
    Configuration conf = new Configuration();
    //检查运行命令
    String[] otherArgs = new GenericOptionsParser(conf,args).getRemainingArgs();
    if(otherArgs.length != 2){
        System.err.println("Usage WordCount <int><out>");
        System.exit(2);
    }
    //配置作业名
    Job job = new Job(conf,"word count");
    //配置作业各个类
    job.setJarByClass(WordCount.class);
    job.setMapperClass(Map.class);
    job.setCombinerClass(Reduce.class);
    job.setReducerClass(Reduce.class);
    job.setOutputKeyClass(Text.class);
    job.setOutputValueClass(IntWritable.class);
    FileInputFormat.addInputPath(job, new Path(otherArgs[0]));
    FileOutputFormat.setOutputPath(job, new Path(otherArgs[1]));
    System.exit(job.waitForCompletion(true) ? 0 : 1);
    }
}
```

步骤 11

在 eclipse 的项目管理器列表中右击项目名 WordCount，选择 export，选择导出为 "JAR file" 格式，如图 8.17 所示。

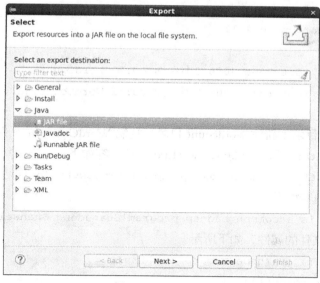

图 8.17 导出窗口

步骤 12

单击 "Next>" 按钮，输入 JAR 包名称 WordCount，如图 8.18 所示。

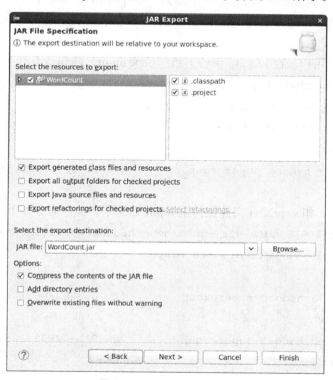

图 8.18 WordCount 打包

步骤 13

单击"Finish"按钮后，生成 JAR 包 WordCount.jar，可以在 /root/workspace/ 目录下查看所生成的 JAR 包，如下所示。

```
[root@node1 ~]# cd /root/workspace/
[root@node1 workspace]# ll
总用量 12
drwxr-xr-x. 4 root root 4096 2月  8 10:59 WordCount
-rw-r--r--. 1 root root 4306 2月  9 00:51 WordCount.jar
```

步骤 14

在 hadoop 用户目录下新建 wordcount 目录，复制 WordCount.jar 文件到该目录下，在该目录下新建文件 swpt1.txt 和 swpt2.txt，swpt1.txt 文件内容如下。

```
This is the first hadoop test program! www.swvtc.cn
```

swpt2.txt 文件内容如下。

```
This program is a common hadoop program!Guangdong Shanwei www.swvtc.cn
```

修改该目录下文件的属性，如下所示。

```
[root@node1 ~]# cd /home/hadoop/wordcount/
[root@node1 wordcount]# ll
总用量 16
-rw-r--r--. 1 hadoop hadoop   55 2月  8 11:44 swpt1.txt
-rw-r--r--. 1 hadoop hadoop   72 2月  8 11:49 swpt2.txt
-rw-r--r--. 1 hadoop hadoop 4306 2月  9 00:55 WordCount.jar
```

步骤 15

切换到 hadoop 用户，在 hadoop 系统上新建目录 input，操作命令如下。

```
[root@node1 wordcount]# su - hadoop
[hadoop@node1 ~]$ hdfs dfs -mkdir /input
[hadoop@node1 ~]$ hdfs dfs -ls /
Found 1 items
drwxr-xr-x   - hadoop supergroup          0 2015-02-08 12:35 /input
```

步骤 16

将文件 swpt1.txt 和 swpt2.txt 文件上传至 input 目录下，操作命令如下。

```
[hadoop@node1 ~]$ hdfs dfs -put /home/hadoop/wordcount/swpt*.txt /input
[hadoop@node1 ~]$ hdfs dfs -ls /input
Found 2 items
-rw-r--r--   3 hadoop supergroup         52 2015-02-08 12:37 /input/swpt1.txt
-rw-r--r--   3 hadoop supergroup         72 2015-02-08 12:37 /input/swpt2.txt
```

步骤 17

运行 WordCount，其中的 "/input" 是输入目录，"/output" 是输出目录，操作命令如下。

```
[hadoop@node1 ~]$ cd wordcount/
[hadoop@node1 wordcount]$ hadoop jar WordCount.jar WordCount /input /output
```

步骤 18

查看输出目录 "/output"。"/output" 目录生成两个新文件，其中 part-r-00000 文件保存 WordCount 的输出结果，如下所示。

```
[hadoop@node1 wordcount]$ hdfs dfs -ls /output
Found 2 items
-rw-r--r--   3 hadoop supergroup          0 2015-02-09 01:02 /output/_SUCCESS
-rw-r--r--   3 hadoop supergroup        118 2015-02-09 01:02 /output/part-r-00000
```

查看 part-r-00000 文件内容，如下所示。

```
[hadoop@node1 wordcount]$ hdfs dfs -text /output/part-r-00000
Guangdong    1
Shanwei 1
This    2
a       1
common  1
first   1
hadoop  2
is      2
program 1
program! 2
test    1
the     1
www.swvtc.cn    2
```

8.4 本章小结

本章完成 hadoop 分布式系统的搭建，介绍 hadoop 的体系架构和 yarn 的架构和 MapReduce 的工作流程，最后使用 Eclipse 完成 MapRecude 项目的开发，实现 hadoop 分布式环境下对文件中出现的单词数目进行统计。

第 9 章 HBase 分布式数据库

HBase 是一个分布式的、面向列的开源数据库，该技术来源于 Fay Chang 所撰写的 Google 论文"Bigtable：一个结构化数据的分布式存储系统"。就像 Bigtable 利用了 Google 文件系统（File System）所提供的分布式数据存储一样，HBase 在 Hadoop 之上提供了类似于 Bigtable 的能力。HBase 是 Apache 的 Hadoop 项目的子项目。HBase 不同于一般的关系数据库，它是一个适于非结构化数据存储的数据库，而且是基于列的而不是基于行的模式。HBase 是一个高可靠性、高性能、面向列、可伸缩的分布式存储系统，利用 HBase 技术可在廉价 PC Server 上搭建起大规模结构化存储集群。

9.1 任务一 HBase 的安装与配置

9.1.1 任务描述

Bigtable 利用 Google 文件系统，HBase 则是利用 Hadoop 的文件系统形成一个可靠的分布式存储系统。安装 Hbase 之前需要了解 Bigtable 概念,了解分布式数据库概念。本节任务是完成 HBase 数据库的安装、配置和使用。

9.1.2 相关知识

1．Bigtable

BigTable 是 Google 设计的分布式数据存储系统，用来处理海量的数据的一种非关系型的数据库。BigTable 是非关系型数据库，是一个稀疏的、分布式的、持久化存储的多维度排序 Map。Bigtable 的设计目的是快速且可靠地处理 PB 级别的数据，并且能够部署到上千台机器上。Bigtable 已经实现了适用性广泛、可扩展、高性能和高可用性。

2．HBase 与 Hadoop 关系

Hadoop 是分布式平台，就把计算和存储都由 Hadoop 自动调节分布到接入的计算机单元中。HBase 是 Hadoop 上实现的数据库，Hadoop 和 HBase 是分布式计算与分布式数据库存储的有效组合。

9.1.3 任务实施

步骤 1

创建用户,分别在 4 台节点机上创建用户 hbase,密码分别为 h1111、h2222、h3333、h4444,uid 和 gid 与 hadoop 用户的相同,设为 uid=660,gid=660。操作分别如下。

```
[root@node1 ~]# useradd -u 660 -g 660 -o hbase
[root@node1 ~]# passwd hbase

[root@node2 ~]# useradd -u 660 -g 660 -o hbase
[root@node2 ~]# passwd hbase

[root@node3 ~]# useradd -u 660 -g 660 -o hbase
[root@node3 ~]# passwd hbase

[root@node4 ~]# useradd -u 660 -g 660 -o hbase
[root@node4 ~]# passwd hbase
```

步骤 2

分别修改 4 台节点机的环境变量。登录 node1 节点机,操作命令如下。

```
[root@node1 ~]# vi /home/hbase/.bash_profile
```

其中的内容:

```
PATH=$PATH:$HOME/bin
```

修改为

```
PATH=$PATH:$HOME/bin:/home/hadoop/bin
```

修改后的内容为

```
# .bash_profile

# Get the aliases and functions
if [ -f ~/.bashrc ]; then
        . ~/.bashrc
fi


# User specific environment and startup programs

PATH=$PATH:$HOME/bin:/home/hadoop/bin

export PATH
```

修改环境变量后,执行以下命令使环境变量设置生效。

```
root@node1 ~]# source /home/hbase/.bash_profile
```

分别登录 node2、node3、node4 节点机修改环境变量,操作同上。

步骤 3

下载 hbase-0.98.10-hadoop2-bin.tar.gz 软件包,地址如下。

http://mirrors.cnnic.cn/apache/hbase/stable/hbase-0.98.10-hadoop2-bin.tar.gz

步骤 4

使用 WinSCP 上传 hbase-0.98.10-hadoop2-bin.tar.gz 软件包到 node1 节点机的 root 目录下,或者直接在 node1 节点机上下载。如图 9.1 所示。

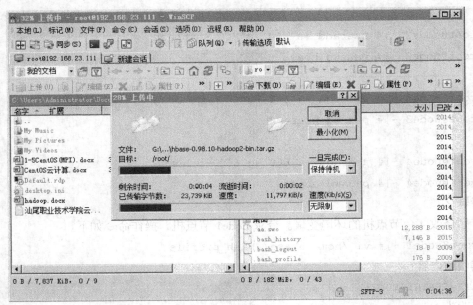

图 9.1 使用 WinSCP 上传 hbase-0.98.10-hadoop2-bin.tar.gz 软件包

步骤 5

解压文件,安装文件。操作命令如下。

```
[root@node1 ~]# cd
[root@node1 ~]# tar xzvf hbase-0.98.10-hadoop2-bin.tar.gz
[root@node1 ~]# cd hbase-0.98.10-hadoop2
[root@node1 hbase-0.98.10-hadoop2]# mv * /home/hbase
```

步骤 6

Hbase 配置文件主要有 hbase-env.sh、hbase-site.xml、regionservers。配置文件在 /home/hbase/conf/目录下。

(1) 修改 hbase-env.sh 文件,主要设置 jdk 路径和设置由 hbase 自动启动 zookeeper。操作命令如下。

```
[root@node1 hbase-0.98.10-hadoop2]# cd /home/hbase
[root@node1 hbase]# vi /home/hbase/conf/hbase-env.sh
```

修改如下两处内容。

```
export JAVA_HOME=/usr/lib/jvm/java-1.7.0
export HBASE_MANAGES_ZK=true
```

（2）修改 hbase-site.xml 文件，将文件中的 <configuration></configuration> 修改为如下内容。

```xml
<configuration>
<property>
<name>hbase.master</name>              #指明 master 节点
<value>node1:60000</value>
</property>
<property>
<name>hbase.master.port</name>
<value>60000</value>
</property>
<property>
<name>hbase.master.maxclockskew</name>
<value>180000</value>
</property>
<property>
<name>hbase.rootdir</name>             #指明数据位置
<value>hdfs://node1:9000/hbase</value> #该值 hdfs://node1:9000 与
#hadoop 的 core-site.xml 配置相同
</property>
<property>
<name>hbase.cluster.distributed</name> #指明是否配置成为集群模式
<value>true</value>
</property>
<property>
<name>hbase.zookeeper.quorum</name>    #指明 zookeeper 安装节点
<value>node2,node3,node4</value>
</property>
<property>
<name>hbase.zookeeper.property.dataDir</name>
                                       #指明 zookeeper 数据存储目录
<value>/home/hbase/tmp/zookeeper</value>
</property>
</configuration>
```

（3）修改配置文件 regionservers，添加 slave 节点的机器名或 IP 地址。操作命令如下。

```
[root@node1 hbase]# vi /home/hbase/conf/regionservers
node2
node3
node4
```

步骤 7

修改 "/home/hbase/" 文件用户主/组属性，操作命令如下。

```
[root@node1 hbase]# chown -R hadoop:hadoop /home/hbase
```

步骤 8

将配置好的 hbase 系统复制到其他 3 个节点机上，操作命令如下。

```
[root@node1 hbase]# cd /home/hbase
[root@node1 hbase]# scp -r * root@node2:/home/hbase
[root@node1 hbase]# scp -r * root@node3:/home/hbase
[root@node1 hbase]# scp -r * root@node4:/home/hbase
```

步骤 9

分别登录 node2、node3、node4 节点机，修改 "/home/hbase/" 文件用户主/组属性。

```
[root@node2 ~]# chown -R hadoop:hadoop /home/hbase
[root@node3 ~]# chown -R hadoop:hadoop /home/hbase
[root@node4 ~]# chown -R hadoop:hadoop /home/hbase
```

9.2 任务二 HBase 管理与 HBase Shell 的使用

9.2.1 任务描述

安装和配置好 HBase 后，需要对 HBase 进行管理和操作，HBase Shell 为用户提供了有用的交互接口。本节任务是掌握 HBase 数据库服务的启动、停止，掌握 HBase Shell 操作命令，掌握使用 HBase Shell 命令创建表，掌握使用 HBase Shell 操作数据表。

9.2.2 相关知识

1．HBase Shell

HBase Shell 为用户提供了一个非常方便的使用方式。HBase Shell 提供了大多数的 HBase 命令，通过 HBase Shell 用户可以方便地创建、删除及修改表，还可以向表中添加数据、列出表中的相关信息等。HBase Shell 的主要命令包括：create 创建表、describe 查看表的结构、enable/disable 表激活/取消、drop 删除表、get/put 表读/写。

2．HBase 基本架构

Hbase 基本架构如图 9.2 所示。

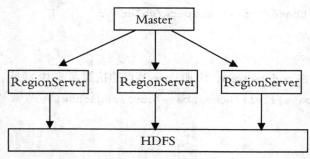

图 9.2 HBase 基本架构

HBase 使用 Zookeeper 来管理集群。在架构层面上分为 Master（Zookeeper 中的 leader）和多个 RegionServer。RegionServer 是调度者，管理 Regions，但是数据是持久化在 HDFS 上的。RegionServer 对应于集群中的一个节点，而一个 RegionServer 负责管理多个 Region。一个 Region 代表一张表的一部分数据，所以在 HBase 中的一张表可能会需要很多个 Region 来存储其数据，但是每个 Region 中的数据并不是杂乱无章的。HBase 在管理 Region 的时候会给每个 Region 定义一个 Rowkey 的范围，落在特定范围内的数据将交给特定的 Region，从而将负载分摊到多个节点上，充分利用分布式的优点。另外，Hbase 会自动调节 Region 所在的位置，如果一个 RegionServer 变得 Hot（大量的请求落在这个 Server 管理的 Region 上），Hbase 就会把 Region 移动到相对空闲的节点，依次保证集群环境被充分利用。

9.2.3 任务实施

步骤 1

启动 hbase 服务。登录 node1 节点机，以用户 hbase 登录或 su – hbase 登录。操作命令如下。

```
[root@node1 hbase]# su - hbase
```

运行脚本，启动服务。操作命令如下。

```
[hadoop@node1 ~]$ start-hbase.sh
```

启动后显示如下信息，表示 hbase 服务启动成功。

```
node3: starting zookeeper, logging to /home/hbase/bin/../logs/hbase-hadoop-zookeeper-node3.out
  node2: starting zookeeper, logging to /home/hbase/bin/../logs/hbase-hadoop-zookeeper-node2.out
  node4: starting zookeeper, logging to /home/hbase/bin/../logs/hbase-hadoop-zookeeper-node4.out
  starting master, logging to /home/hbase/bin/../logs/hbase-hadoop-master-node1.out
  node2: starting regionserver, logging to /home/hbase/bin/../logs/hbase-hadoop-regionserver-node2.out
  node3: starting regionserver, logging to /home/hbase/bin/../logs/hbase-hadoop-regionserver-node3.out
  node4: starting regionserver, logging to /home/hbase/bin/../logs/hbase-hadoop-regionserver-node4.out
```

步骤 2

检查运行结果，Master 节点机的进程如下所示。

```
[hadoop@node1 ~]$ jps
36059 ResourceManager
42439 HMaster
35834 SecondaryNameNode
```

```
42584 Jps
35633 NameNode
```

Slave 节点机的进程如下所示。

```
[hadoop@node2 ~]$ jps        [hadoop@node3 ~]$ jps        [hadoop@node4 ~]$ jps
41001 HRegionServer          37245 DataNode               41172 Jps
40652 HQuorumPeer            40636 Jps                    40077 HQuorumPeer
38030 NodeManager            39895 HQuorumPeer            41103 HRegionServer
37855 DataNode               37417 NodeManager            37596 NodeManager
41072 Jps                    40575 HRegionServer          37424 DataNode
```

注意：4 台节点机时间要基本一致，否则 slave 节点机 HRegionServer 服务会自动停止。

步骤 3

停止 hbase 服务，操作命令如下。

```
[hadoop@node1 ~]$ stop-hbase.sh
stopping hbase
node2: stopping zookeeper.
node4: stopping zookeeper.
node3: stopping zookeeper.
```

步骤 4

登录 hbase 的 web 界面，如图 9.3 所示。

图 9.3 hbase 的 web 界面

步骤 5

启动 hbase shell，如下所示。

```
[hadoop@node1 ~]$ hbase shell
HBase Shell; enter 'help<RETURN>' for list of supported commands.
Type "exit<RETURN>" to leave the HBase Shell
Version 0.98.10-hadoop2, r96878ece501b0643e879254645d7f3a40eaf101f, Mon Dec 15 23:00:20 PST 2014

hbase(main):001:0>
```

步骤 6

建立表 scores，两个列簇：grade 和 course，操作命令如下。

```
hbase(main):001:0> create 'scores','grade','course'
0 row(s) in 2.9470 seconds

=> Hbase::Table - scores
```

步骤 7

查看当前 hbase 中的表，操作命令如下。

```
hbase(main):002:0> list
TABLE
scores
1 row(s) in 0.0360 seconds

=> ["scores"]
```

步骤 8

添加 6 条记录。

（1）记录 1：jie，grade：143cloud。操作命令如下。

```
hbase(main):003:0> put 'scores','jie','grade:','143cloud'
0 row(s) in 0.2420 seconds
```

（2）记录 2：jie，course：math，85。操作命令如下。

```
hbase(main):004:0> put 'scores','jie','course:math','85'
0 row(s) in 0.0560 seconds
```

（3）记录 3：jie，course：cloud，92。操作命令如下。

```
hbase(main):005:0> put 'scores','jie','course:cloud','92'
0 row(s) in 0.0330 seconds
```

（4）记录 4：shi，grade：133soft。操作命令如下。

```
hbase(main):006:0> put 'scores','shi','grade:','133soft'
0 row(s) in 0.0340 seconds
```

（5）记录 5：shi，grade：math，86。操作命令如下。

```
hbase(main):007:0> put 'scores','shi','course:math','86'
0 row(s) in 0.0580 seconds
```

(6) 记录6: shi, grade: cloud, 90。操作命令如下。

```
hbase(main):008:0> put 'scores','shi','course:cloud','90'
0 row(s) in 0.0320 seconds
```

(1) 读取 jie 的记录，操作命令如下。

```
hbase(main):009:0> get 'scores','jie'
COLUMN                  CELL
 course:cloud           timestamp=1422552658530, value=92
 course:math            timestamp=1422552591848, value=85
 grade:                 timestamp=1422552542657, value=143cloud
3 row(s) in 0.0830 seconds
```

(2) 读取 jie 的班级，操作命令如下。

```
hbase(main):010:0> get 'scores','jie','grade'
COLUMN                  CELL
 grade:                 timestamp=1422552542657, value=143cloud
1 row(s) in 0.0290 seconds
```

步骤 10

查看整个表记录，操作命令如下。

```
hbase(main):011:0> scan 'scores'
ROW                     COLUMN+CELL
 jie                    column=course:cloud, timestamp=1422552658530, value=92
 jie                    column=course:math, timestamp=1422552591848, value=85
 jie                    column=grade:, timestamp=1422552542657, value=143cloud
 shi                    column=course:cloud, timestamp=1422553176842, value=90
 shi                    column=course:math, timestamp=1422553089886, value=86
 shi                    column=grade:, timestamp=1422552722718, value=133soft
2 row(s) in 0.0630 seconds
```

步骤 11

按列查看表记录，操作命令如下。

```
hbase(main):012:0> scan 'scores',{COLUMNS=>'course'}
ROW                     COLUMN+CELL
 jie                    column=course:cloud, timestamp=1422552658530, value=92
 jie                    column=course:math, timestamp=1422552591848, value=85
 shi                    column=course:cloud, timestamp=1422553176842, value=90
 shi                    column=course:math, timestamp=1422553089886, value=86
2 row(s) in 0.0530 seconds
```

步骤 12

删除指定记录，操作命令如下。

```
hbase(main):013:0> delete 'scores','shi','grade'
0 row(s) in 0.0670 seconds
```

删除后，执行 scan 命令，显示如下信息。

```
hbase(main):014:0> scan 'scores'
ROW                   COLUMN+CELL
 jie                  column=course:cloud, timestamp=1422552658530, value=92
 jie                  column=course:math, timestamp=1422552591848, value=85
 jie                  column=grade:, timestamp=1422552542657, value=143cloud
 shi                  column=course:cloud, timestamp=1422553176842, value=90
 shi                  column=course:math, timestamp=1422553089886, value=86
2 row(s) in 0.0610 seconds
```

步骤 13

增加新的列簇，操作命令如下。

```
hbase(main):025:0> alter 'scores',NAME=>'age'
Updating all regions with the new schema...
0/1 regions updated.
1/1 regions updated.
Done.
0 row(s) in 2.3980 seconds
```

步骤 14

查看表结构，操作命令如下。

```
hbase(main):026:0> describe 'scores'
Table scores is ENABLED
COLUMN FAMILIES DESCRIPTION
{NAME => 'age', DATA_BLOCK_ENCODING => 'NONE', BLOOMFILTER => 'ROW', REPLICATION
_SCOPE => '0', COMPRESSION => 'NONE', VERSIONS => '1', TTL => 'FOREVER', MIN_VER
SIONS => '0', KEEP_DELETED_CELLS => 'FALSE', BLOCKSIZE => '65536', IN_MEMORY =>
'false', BLOCKCACHE => 'true'}
{NAME => 'course', DATA_BLOCK_ENCODING => 'NONE', BLOOMFILTER => 'ROW', REPLICAT
ION_SCOPE => '0', VERSIONS => '1', COMPRESSION => 'NONE', MIN_VERSIONS =>
'0', T
TL => 'FOREVER', KEEP_DELETED_CELLS => 'FALSE', BLOCKSIZE => '65536', IN_MEMORY
 => 'false', BLOCKCACHE => 'true'}
{NAME => 'grade', DATA_BLOCK_ENCODING => 'NONE', BLOOMFILTER => 'ROW',
```

```
REPLICATI
   ON_SCOPE => '0', VERSIONS => '1', COMPRESSION => 'NONE', MIN_VERSIONS => '0', TT
   L => 'FOREVER', KEEP_DELETED_CELLS => 'FALSE', BLOCKSIZE => '65536',
IN_MEMORY => 'false', BLOCKCACHE => 'true'}
   3 row(s) in 0.0970 seconds
```

步骤 15

删除列簇,操作命令如下。

```
hbase(main):028:0> alter 'scores',NAME=>'age',METHOD=>'delete'
Updating all regions with the new schema...
0/1 regions updated.
1/1 regions updated.
Done.
0 row(s) in 2.3980 seconds
```

步骤 16

删除表,操作命令如下。

```
hbase(main):015:0> disable 'scores'
0 row(s) in 1.6400 seconds
hbase(main):016:0> drop 'scores'
0 row(s) in 0.5400 seconds
```

9.3 本章小结

本章主要介绍 HBase 的基本架构,介绍了 HBase 的安装和配置,HBase 是一个高可靠性、高性能、面向列、可伸缩的分布式存储系统,一个开源的数据库。HBase Shell 提供了一个有效的交互接口,HBase 的 web 界面方便了用户操作。

第 10 章 Storm 环境的搭建与管理

在过去十年里，随着互联网应用的高速发展，企业积累的数据量越来越大，越来越多。随着 Google MapReduce、Hadoop 等相关技术的出现，处理大规模数据变得简单起来，但是这些数据处理技术都不是实时的系统，它们的设计目标也不是实时计算。毕竟实时的计算系统和基于批处理模型的系统（如 Hadoop）有着本质的区别。

但是随着大数据业务的快速增长，针对大规模数据处理的实时计算变成了一种业务上的需求，缺少"实时的 Hadoop 系统"已经成为整个大数据生态系统中的一个巨大缺失。Storm 正是在这样的需求背景下出现的，Storm 很好地满足了这一需求。

Storm 是一个开源的分布式实时计算系统，可以简单、可靠、高效地处理大量数据流，支持水平扩展且适用场景广，具有实时性、高容错性、健壮性、扩展性、语言无关性等优点。当 Storm 在进行数据处理时，会以很高的处理速度将中间结果全部写入到内存中。一般来说，在一个小的 Storm 集群里，每个结点以达到近百万消息每秒的速度进行计算。Storm 与 Hadoop 在处理数据时还有一个很大的不同，那就是 Storm 是通过创建拓扑结构来处理没有终点的数据流，而这些转换工作会一直进行，并持续处理数据流中新到达的数据。

表 10.1 列举了 Storm 和 Hadoop 的基本信息对比情况，通过该表方便读者快速了解 Storm 的一些基本概念。

表 10.1　Storm 和 Hadoop 的基本信息对比

系统名称	Hadoop	Storm
系统角色	JobTracke	Nimbus
	TaskTracker	Supervisor
	Child	Worker
应用名称	Job	Topology
组件接口	Mapper/Reducer	Spout/Bolt

Storm 的主要特点如下。

（1）简单的编程模型。类似于 MapReduce 降低了并行批处理复杂性，Storm 降低了进行实时处理的复杂性。

（2）可以使用各种编程语言。你可以在 Storm 之上使用各种编程语言。默认支持 Clojure、Java、Ruby 和 Python。要增加对其他语言的支持，只需实现一个简单的 Storm 通信协议即可。

（3）容错性。Storm 会管理工作进程和节点的故障。

（4）水平扩展。计算是在多个线程、进程和服务器之间并行进行的。

（5）可靠的消息处理。Storm 保证每个消息至少能得到一次完整处理。任务失败时，它会负责从消息源重试消息。

（6）快速。系统的设计保证了消息能得到快速的处理，使用 ZeroMQ 作为其底层消息队列。

（7）本地模式。Storm 有一个"本地模式"，可以在处理过程中完全模拟 Storm 集群。这让你可以快速进行开发和单元测试。

10.1 任务一　Storm 的安装与配置

10.1.1 任务描述

本节任务是完成 Storm 的安装与部署，包括安装相关的依赖软件（Python、JDK、gcc-c++、uuid*、libtool、libuuid、libuuid-devel 等）及安装运行 Storm 必备的工具包（ZeroMQ、JZMQ、Zookeeper）。

10.1.2 相关知识

1. Maven

Maven 是一个完全采用 Java 语言编写的开源项目管理工具。它通过项目对象模型（project object model，POM）来管理项目，所有的项目配置信息都被定义在一个叫做 pom.xml 的文件中，通过该文件，Maven 可以管理项目的整个声明周期，包括编译、构建、测试、发布、报告等。

- 目前 Apache 基金会下的绝大多数项目都是采用 Maven 进行管理。
- Maven 本身还支持多种插件，可以方便灵活地控制项目。
- Maven 的核心配置文件是 pom.xml，该文件包含了如何构建项目的大多数配置信息。

2. Git

Git 在维基百科上的定义是一个免费的、分布式的版本控制工具，或是一个强调了速度快的源代码管理工具。它是 Linux Torvalds 为了帮助管理 Linux 内核开发而开发的一个开源的分布式版本控制软件，用以有效、高速地进行从很小到非常大的项目版本管理。每一个 Git 的工作目录都是一个完全独立的代码库，并拥有完整的历史记录和版本追踪能力，不依赖于网络和中心服务器。由于 Git 对分支具有良好的控制，目前，许多开发者和开源项目都习惯使用它来进行各分支管理。

- git clone: 这是较为简单的一种初始化方式，当远程已经存在一个的 Git 版本库，只需通过该命令即可在本地复制一份。
- Git 能够提供方便的本地分支等特性，是因为 Git 存储版本控制信息时使用它自己定义的一套文件系统存储机制。

3. Python

Python 是一种面向对象、直译式计算机程序设计语言，也是一种功能强大而完善的通用型语言，有着"胶水语言"的昵称，发展至今已有十多年的历史，成熟且稳定。

- Python 语法简洁而清晰，具有丰富和强大的类库。
- 广泛应用于处理系统管理任务和 Web 编程。
- 适合完成各种高层任务，几乎可以在所有的操作系统中运行。

4. ZeroMQ

ZeroMQ，简称 ZMQ，是一个简单好用的传输层，像框架一样的一个 socket library，它使得 Socket 编程更加简单、简洁和性能更高。它是一个消息处理队列库，可在多个线程、内核和主机盒之间弹性伸缩。ZMQ 并不像是一个传统意义上的消息队列服务器，事实上，它也根本不是一个服务器，它更像是一个底层的网络通信库，在 Socket API 之上做了一层封装，将网络通信、进程通信和线程通信抽象为统一的 API 接口。

- ZMQ 的明确目标是"成为标准网络协议栈的一部分，之后进入 Linux 内核"。
- ZMQ 用于 node 与 node 间的通信，node 可以是主机或者是进程。

5. JZMQ

JZMQ 是 ZeroMQ 的 Java Binding，是使用 Maven 进行部署和管理的。

6. ZooKeeper

（1）ZooKeeper 简介。

ZooKeeper 是一个开放源码的分布式应用程序协调服务，是 Google 的 Chubby 一个开源的实现。它是一个为分布式应用提供一致性服务的软件，提供的功能包括配置维护、名字服务、分布式同步、组服务等。

（2）ZooKeeper 功能。

Storm 中使用 Zookeeper 主要用于 Storm 集群各节点的分布式协调工作，具体功能如下。

- 存储客户端提供的 topology 任务信息，nimbus 负责将任务分配信息写入 Zookeeper，supervisor 从 Zookeeper 上读取任务分配信息。
- 存储 supervisor 和 worker 的心跳（包括它们的状态），使得 nimbus 可以监控整个集群的状态，从而重启一些挂掉的 worker。
- 存储整个集群的所有状态信息和配置信息。

10.1.3 任务实施

本节任务需要 4 台节点机组成集群，每个节点机上安装 CentOS-6.5-x86_64 系统。4 台节点机使用的 IP 地址分别为 192.168.23.111、192.168.23.112、192.168.23.113、192.168.23.114，对应节点主机名为 node1、node2、node3、node4。节点机 node1 作为主节点机，其他作为从节点机。

步骤 1

关闭安全设置，分别登录 node1、node2、node3、node4 节点机，关闭 selinux 和防火墙。操作命令如下。

（1）关闭 selinux，操作命令如下。

```
[root@node1 ~]# setenforce 0
[root@node1 ~]# vi /etc/selinux/config
```

将"/etc/selinux/config"配置文件中 SELINUX=enforcing 改为 SELINUX=disabled。

（2）关闭防火墙，操作命令如下。

```
[root@node1 ~]# service iptables stop
[root@node1 ~]# chkconfig --level 2345 iptables off
```

其他节点机操作相同。

步骤 2

登录 node1 节点机安装 jdk、gcc、gcc-c++ 、make、cmake、openssl-devel、ncurses-devel、uuid、libuuid、libuuid-devel、libtool。其中：jdk 要求每个节点机都安装。

（1）安装 jdk，操作命令如下。

```
[root@node1~]#cd /media/CentOS_6.5_Final/Packages/
[root@localhost Packages]#rpm -ivh java-1.7.0-openjdk-devel-1.7.0.45-2.4.3.3.el6.x86_64.rpm
```

（2）安装 gcc 开发包，操作命令如下。

```
[root@node1 Packages]#rpm -ivh mpfr-2.4.1-6.el6.x86_64.rpm
[root@node1 Packages]#rpm -ivh cpp-4.4.7-4.el6.x86_64.rpm
[root@node1 Packages]#rpm -ivh ppl-0.10.2-11.el6.x86_64.rpm
[root@node1 Packages]#rpm -ivh cloog-ppl-0.15.7-1.2.el6.x86_64.rpm
[root@node1 Packages]#rpm -ivh gcc-4.4.7-4.el6.x86_64.rpm
```

（3）安装 gcc-c++开发包，操作命令如下。

```
[root@node1 Packages]#rpm -ivh libstdc++-devel-4.4.7-4.el6.x86_64.rpm
[root@node1 Packages]#rpm -ivh gcc-c++-4.4.7-4.el6.x86_64.rpm
```

（4）安装 gcc-gfortran 开发包，操作命令如下。

```
[root@node1 Packages]#rpm -ivh libgfortran-4.4.7-4.el6.x86_64.rpm
[root@node1 Packages]#rpm -ivh gcc-gfortran-4.4.7-4.el6.x86_64.rpm
```

（5）安装 camke，操作命令如下。

```
[root@node1 Packages]#rpm -ivh cmake-2.6.4-5.el6.x86_64.rpm
```

（6）安装 openssl-devel，操作命令如下。

```
[root@node1Packages]# rpm -ivh keyutils-libs-devel-1.4-4.el6.x86_64.rpm
[root@node1 Packages]# rpm -ivh libcom_err-devel-1.41.12-18.el6.x86_64.rpm
[root@node1 Packages]# rpm -ivh libsepol-devel-2.0.41-4.el6.x86_64.rpm
[root@node1 Packages]# rpm -ivh libselinux-devel-2.0.94-5.3.el6_4.1.x86_64.rpm
[root@node1 Packages]# rpm -ivh krb5-devel-1.10.3-10.el6_4.6.x86_64.rpm
[root@node1 Packages]# rpm -ivh zlib-devel-1.2.3-29.el6.x86_64.rpm
[root@node1 Packages]# rpm -ivh openssl-devel-1.0.1e-15.el6.x86_64.rpm
```

（7）安装 ncurses-devel，操作命令如下。

[root@localhost Packages]# rpm -ivh ncurses-devel-5.7-3.20090208.el6.x86_64.rpm

（8）安装 uuid*，操作命令如下。

[root@node1 Packages]# rpmuuid-1.6.1-10.el6.x86_64.rpm

[root@node1 Packages]# rpmrpm -ivh uuidd-2.17.2-12.14.el6.x86_64.rpm

（9）安装 libuuid 和 libuuid-devel，操作命令如下。

[root@node1 Packages]#rpm -ivh libuuid-2.17.2-12.14.el6.x86_64.rpm

[root@node1 Packages]#rpm -ivh libuuid-devel-2.17.2-12.14.el6.x86_64.rpm

（10）安装 libtool，操作命令如下。

[root@node1 Packages]#rpm -ivh autoconf-2.63-5.1.el6.noarch.rpm

[root@node1 Packages]#rpm -ivh automake-1.11.1-4.el6.noarch.rpm

[root@node1 Packages]#rpm -ivh libtool-2.2.6-15.5.el6.x86_64.rpm

另外，如果能连上因特网，则可以使用 yum 安装开发包。安装命令如下。

```
yum install java-1.7.0-openjdk*
yum install gcc
yum install gcc-c++
yum install make
yum install cmake
yum install openssl-devel
yum install ncurses-devel
yum install uuid*
yum install libtool
yum install libuuid
yum install libuuid-devel
```

步骤 3

（1）下载 maven 软件包，地址：

http://mirror.bit.edu.cn/apache/maven/maven-3/3.2.5/binaries/apache-maven-3.2.5-bin.tar.gz

（2）解压 maven 软件包。

[root@node1 Packages]#cd

[root@node1~]# tar xvzf pache-maven-3.2.5-bin.tar.gz

（3）把 maven 软件移到目录 /usr/local 下，命令如下。

[root@node1~]# mv pache-maven-3.2.5 /usr/local/maven

（4）编辑环境变量。

[root@node1 ~]#vi /etc/profile

添加内容如下。

export M2_HOME=/usr/local/maven

```
export M2=$M2_HOME/bin
export MAVEN_OPTS="-Xms256m -Xmx512m"
export PATH="$M2:$PATH"
```

（5）使环境变量生效。

```
[root@node1 ~]# source /etc/profile
```

（6）检查maven版本，显示如下表示成功。

```
[root@node1 ~]# mvn -version
Apache Maven 3.2.5 (12a6b3acb947671f09b81f49094c53f426d8cea1;
2014-12-15T01:29:23+08:00)
Maven home: /usr/local/maven
Java version: 1.7.0_45, vendor: Oracle Corporation
Java home: /usr/lib/jvm/java-1.7.0-openjdk-1.7.0.45.x86_64/jre
Default locale: zh_CN, platform encoding: UTF-8
OS name: "linux", version: "2.6.32-431.el6.x86_64", arch: "amd64", family:
"unix"
```

（7）修改配置文件。

```
[root@node1 ~]# vi /usr/local/maven/conf/settings.xml
```

在<mirrors></mirrors>内添加如下内容，其他的不需改动。

```
<mirror>
<id>nexus-osc</id>
<mirrorOf>*</mirrorOf>
<name>Nexusosc</name>
<url>http://maven.oschina.net/content/groups/public/</url>
</mirror>
```

在<profiles></profiles>内添加如下内容。

```
<profile>
<id>jdk-1.7</id>
<activation>
<jdk>1.7</jdk>
</activation>`
<repositories>
<repository>
<id>nexus</id>
<name>local private nexus</name>
<url>http://maven.oschina.net/content/groups/public/</url>
<releases>
<enabled>true</enabled>
</releases>
```

```xml
<snapshots>
<enabled>false</enabled>
</snapshots>
</repository>
</repositories>
<pluginRepositories>
<pluginRepository>
<id>nexus</id>
<name>local private nexus</name>
<url>http://maven.oschina.net/content/groups/public/</url>
<releases>
<enabled>true</enabled>
</releases>
<snapshots>
<enabled>false</enabled>
</snapshots>
</pluginRepository>
</pluginRepositories>
</profile>
```

步骤 4

安装 libcurl-devel,操作命令如下。

```
[root@node1 ~]# cd /media/CentOS_6.5_Final/Packages/
[root@node1 Packages]# rpm -ivh libcurl-7.19.7-37.el6_4.x86_64.rpm
[root@node1 Packages]# rpm -ivh libidn-devel-1.18-2.el6.x86_64.rpm
[root@node1 Packages]# rpm -ivh libcurl-devel-7.19.7-37.el6_4.x86_64.rpm
```

步骤 5

如果不需要使用 git 下载软件,这步可以跳过。

(1) 下载 git 软件包,地址:

https://www.kernel.org/pub/software/scm/git/git-2.3.0.tar.gz

(2) 解压。

```
[root@node1 Packages]# cd
[root@node1 ~]# tar xvzf git-2.3.0.tar.gz
```

(3) 配置。

```
[root@node1 ~]# cd git-2.3.0
[root@node1 git-2.3.0]# ./configure--with-curl --with-expat
```

(4) 编译。

```
[root@node1 git-2.3.0]# make
```

（5）安装。

```
[root@node1 git-2.3.0]# make install
```

（6）检查 git 版本。

```
[root@node1 git-2.3.0]# git --version
git version 2.3.0
```

步骤 6

安装 python 可以下载最新版本安装，也可以使用系统光盘安装。

使用系统盘安装。分别登录 node1、node2、node3、node4 节点机，安装 python。

（1）安装命令如下。

```
[root@node1 ~]# cd /media/CentOS_6.5_Final/Packages/
[root@node1 Packages]# rpm -ivh python-2.6.6-51.el6.x86_64.rpm
[root@node1 Packages]# rpm -ivh python-devel-2.6.6-51.el6.x86_64.rpm
```

（2）运行 python，操作命令如下。

```
[root@node1 Packages]# python
Python 2.6.6 (r266:84292, Nov 22 2013, 12:16:22)
[GCC 4.4.7 20120313 (Red Hat 4.4.7-4)] on linux2
Type "help", "copyright", "credits" or "license" for more information.
>>>
```

下载并安装最新版本。

（1）下载 python 软件包，地址：

https://www.python.org/ftp/python/2.7.9/Python-2.7.9.tgz

（2）解压。

```
[root@node1 Packages]# cd
[root@node1 ~]# tar xvzf Python-2.7.9.tgz
```

（3）配置。

```
[root@node1 ~]# cd Python-2.7.9
[root@node1 Python-2.7.9]# ./configure --prefix=/usr/local/python
```

（4）编译。

```
[root@node1 Python-2.7.9]# make
```

（5）安装。

```
[root@node1 Python-2.7.9]# make install
```

（6）设置环境变量。

```
[root@node1 Python-2.7.9]# vi /etc/profile
```

添加如下内容。

```
export PYTHON_HOME=/usr/local/python
export PATH=$PYTHON_HOME/bin:$PATH
```

（7）使环境变量生效。

```
[root@node1 Python-2.7.9]# source /etc/profile
```

（8）运行 python，操作命令如下。

```
[root@node1 Python-2.7.9]#python
Python 2.7.9 (default, Feb 18 2015, 13:15:51)
[GCC 4.4.7 20120313 (Red Hat 4.4.7-4)] on linux2
Type "help", "copyright", "credits" or "license" for more information.
>>>
```

（9）复制 python 到其他节点。

将 node1 节点机的"/usr/local/python"复制到其他 3 台节点机上，操作如下。

```
[root@node1 Python-2.7.9]# cd
[root@node1 ~]# scp -r -p /usr/local/python root@node2:/usr/local/
[root@node1 ~]# scp -r -p /usr/local/python root@node3:/usr/local/
[root@node1 ~]# scp -r -p /usr/local/python root@node4:/usr/local/
```

（10）修改 node2、node3、node4 节点机的环境变量，与（6）、（7）相同。

步骤 7

（1）下载 ZeroMQ 软件包，地址：

http://download.zeromq.org/zeromq-4.0.5.tar.gz

（2）解压。

```
[root@node1 ~]# cd
[root@node1 ~]# tar xvzf zeromq-4.0.5.tar.gz
```

（3）配置。

```
[root@node1 ~]# cd zeromq-4.0.5
[root@node1 zeromq-4.0.5]# ./configure --prefix=/home/local/zeromq
```

（4）编译。

```
[root@node1 zeromq-4.0.5]# make
```

（5）安装。

```
[root@node1 zeromq-4.0.5]# make install
```

（6）更新动态链接库。

```
[root@node1 zeromq-4.0.5]# ldconfig
```

（7）设置环境变量。

```
[root@node1 zeromq-4.0.5]# vi /etc/profile
```

添加如下内容。

```
export CPPFLAGS=-I/home/local/zeromq/include/
export LDFLAGS=-L/home/local/zeromq/lib/
```

（8）使环境变量生效。

```
[root@node1 zeromq-4.0.5]# source /etc/profile
```

步骤 8

（1）下载 JZMQ 软件包，地址：

```
[root@node1 zeromq-4.0.5]# cd
```

```
[root@node1 ~]# wget https://codeload.github.com/zeromq/jzmq/zip/master
```
（2）解压文件。
```
[root@node1 ~]# unzip master
```
（3）配置。
```
[root@node1 ~]# cd jzmq-master/
[root@node1 jzmq-master]# ./autogen.sh
[root@node1 jzmq-master]# ./configure --prefix=/home/local/jzmq
```
（4）编译。
```
[root@node1 jzmq-master]# make
```
（5）安装。
```
[root@node1 jzmq-master]# make install
```
（6）设置环境变量。
```
[root@node1 jzmq-master]# vi /etc/profile
```
添加如下内容。
```
export LD_LIBRARY_PATH=$LD_LIBRARY_PATH:/home/local/zeromq/lib/:/home/local/jzmq/lib/
```
（7）使环境变量生效。
```
[root@node1 jzmq-master]# source /etc/profile
```

步骤9

（1）下载 Zookeeper 软件包，地址：
http://mirror.bjtu.edu.cn/apache/zookeeper/zookeeper-3.4.6/zookeeper-3.4.6.tar.gz

（2）解压安装。
```
[root@node1 jzmq-master]# cd
[root@node1 ~]# tar xvzf zookeeper-3.4.6.tar.gz
[root@node1 ~]# mv zookeeper-3.4.6 /home/local/zookeeper
```
（3）修改用户属性。
```
[root@node1 ~]# chown -R root:root /home/local/zookeeper/
```
（4）设置环境变量。
```
[root@node1 ~]# vi /etc/profile
```
添加如下内容。
```
export ZOOKEEPER_HOME=/home/local/zookeeper
export PATH=$PATH:$ZOOKEEPER_HOME/bin
```
（5）使环境变量生效。
```
[root@node1 ~]# source /etc/profile
```
（6）参数配置。
```
[root@node1 ~]# cd /home/local/zookeeper/conf
[root@node1 conf]# mv zoo_sample.cfg zoo.cfg
```

```
[root@node1 conf]# vi zoo.cfg
```
配置文件添加如下内容。
```
dataDir=/home/local/zookeeper/data
dataLogDir=/home/local/zookeeper/log
clientPort=2181
server.1=node1:2888:3888
server.2=node2:2888:3888
server.3=node3:2888:3888
server.4=node4:2888:3888
```
其中，dataDir 指定 Zookeeper 的数据文件目录；server.id=host:port1:port2，id 是为每个 Zookeeper 节点的编号，保存在 dataDir 目录下的 myid 文件中，node1～node4 表示各个 Zookeeper 节点的 hostname，第一个 port（2888：代表集群内通信）是用于连接 leader 的端口，第二个 port（3888：代表集群外通信）是用于 leader 选举的端口。

修改 zoo.cfg 文件，启用日志自动清理功能。如下所示。
```
# The number of snapshots to retain in dataDir
autopurge.snapRetainCount=3
# Purge task interval in hours
# Set to "0" to disable auto purge feature
autopurge.purgeInterval=24
```
注意事项：Zookeeper 运行过程中会在 dataDir 目录下生成很多日志和快照文件，而 Zookeeper 运行进程并不负责定期清理合并这些文件，导致占用大量磁盘空间，因此，需要定期清除没用的日志和快照文件。

（7）在$ZOOKEEPER_HOME 目录下新建两目录。
```
[root@node1 conf]# mkdir /home/local/zookeeper/{data,log}
```
（8）在$ZOOKEEPER_HOME/data 目录下创建一个文件：myid。myid 的内容为上面配置的 server.<id>中 id 数字，不同节点机 myid 的内容不同。
```
[root@node1 conf]# cd /home/local/zookeeper/data
[root@node1 data]# vi myid
```
添加内容为 1。

步骤 10

（1）下载 Storm 软件包。地址：
http://mirror.bjtu.edu.cn/apache/storm/apache-storm-0.9.3/apache-storm-0.9.3.tar.gz

（2）解压安装。
```
[root@node1 data]# cd
[root@node1 ~]# tar xvzf apache-storm-0.9.3.tar.gz
[root@node1 ~]# mv apache-storm-0.9.3 /home/local/storm
```
（3）修改用户属性。

```
[root@node1 ~]# chown -R root:root /home/local/storm/
```
（4）设置环境变量。
```
[root@node1 ~]# vi /etc/profile
```
文件添加如下内容。
```
export STORM_HOME=/home/local/storm
export PATH=$PATH:$STORM_HOME/bin
```
（5）使环境变量生效。
```
[root@node1 ~]# source /etc/profile
```
（6）参数配置。
```
[root@node1 ~]# cd /home/local/storm/conf
[root@node1 conf]# vi storm.yaml
```
配置文件添加如下内容。
```
storm.zookeeper.servers:
    - "node1"
    - "node2"
    - "node3"
    - "node4"
nimbus.host: "node1"
storm.local.dir: "/home/local/storm/temp"
storm.zookeeper.port: 2181
supervisor.slots.ports:
 - 6700
 - 6701
 - 6702
 - 6703
```

这里要注意：格式的要求每一项的开始时要加空格，冒号后也必须要加空格。

supervisor.slots.ports 对于每一台工作机器，这个配置指定在这台工作机器上运行多少工作进程，每个进程使用一个独立端口来接收消息，这个配置同时也指定使用哪些端口。

（7）在$STORM_HOME 目录下新建目录 temp。
```
[root@node1 conf]# mkdir /home/local/storm/temp
```

步骤 11

将 node1 节点机 "/home/local/" 目录下文件复制到其他 3 台节点机上，操作命令如下。
```
[root@node1 ~]# scp -r /home/local root@node2:/home/local
[root@node1 ~]# scp -r /home/local root@node3:/home/local
[root@node1 ~]# scp -r /home/local root@node4:/home/local
```

步骤 12

分别登录 node2、node3 和 node4 节点机修改环境变量，操作命令如下。
```
[root@node2 ~]# vi /etc/profile
```

添加如下内容。

```
export CPPFLAGS=-I/home/local/zeromq/include/
export LDFLAGS=-L/home/local/zeromq/lib/
 export LD_LIBRARY_PATH=$LD_LIBRARY_PATH:/home/local/zeromq/lib/:/home/local/jzmq/lib/
export ZOOKEEPER_HOME=/home/local/zookeeper
export STORM_HOME=/home/local/storm
export PATH=$PATH:$ZOOKEEPER_HOME/bin:$STORM_HOME/bin
```

使用环境变量生效：

```
[root@node2 ~]# source /etc/profile
```

其他两台一样操作。

步骤 13

分别登录 node1、node2、node3、node4 节点机，修改$ZOOKEEPER_HOME/data/myid 的内容，按照配置文件的 server.<id>中 id 的数字，修改的内容如下。

node2 的值为 2，node3 的值为 3，node4 的值为 4

10.2 任务二　Storm 的管理

10.2.1　任务描述

Storm 系统搭建完成后，需要启动相关服务，检查运行状态。本节任务主要完成 Storm 服务和 Zookeeper 服务的启动、停止和日常监控操作，练习 Storm 客户端操作命令。

10.2.2　相关知识

Storm 集群由一个主节点和多个工作节点组成。主节点运行了一个名为"Nimbus"的守护进程，用于分配代码、布置任务及故障检测。每个工作节点都运行了一个名为"Supervisor"的守护进程，用于监听工作，开始并终止工作进程。Nimbus 和 Supervisor 都能快速失败，而且是无状态的，这样一来它们就变得十分健壮，两者的协调工作是由 Apache ZooKeeper 来完成的。

1．ZooKeeper 工作原理

Zookeeper 的核心是原子广播，这个机制保证了各个 Server 之间的同步。实现这个机制的协议叫做 Zab 协议。Zab 协议有两种模式，分别是恢复模式（选主）和广播模式（同步）。当服务启动或者在领导者崩溃后，Zab 就进入了恢复模式。当领导者被选举出来，且大多数 Server 完成了和 leader 的状态同步以后，恢复模式就结束了。状态同步保证了 leader 和 Server 具有相同的系统状态。每个 Server 在工作过程中有 3 种状态。

（1）LOOKING：当前 Server 不知道 leader 是谁，正在搜寻。

（2）LEADING：当前 Server 即为选举出来的 leader。

（3）FOLLOWING：leader 已经选举出来，当前 Server 与之同步。

Zookeeper 中的角色主要有以下 3 类，如表 10.2 所示。

表 10.2 Zookeeper 角色分析

角色		描述
领导者（Leader）		领导者负责进行投票的发起和决议，更新系统状态
学习者（Learner）	跟随者（Follower）	Follower 用于接收客户请求并向客户端返回结果，在选主过程中参与投票
	观察者（ObServer）	ObServer 可以接收客户端连接，将写请求转发给 leader 节点。但 ObServer 不参加投票过程，只同步 leader 的状态。ObServer 的目的是扩展系统，提高读取速度
客户端（Client）		请求发起方

Zookeeper 系统模型如图 10.1 所示。

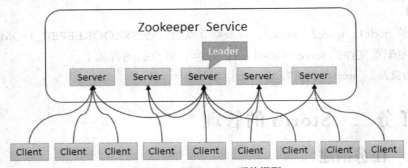

图 10.1 Zookeeper 系统模型

2．Storm 集群组件

Storm 集群中包含两类节点：主控节点（Master Node）和工作节点（Work Node）。其分别对应的角色如下。

主控节点（Master Node）上运行一个被称为 Nimbus 的后台程序，它负责在 Storm 集群内分发代码，分配任务给工作机器，并且负责监控集群运行状态。Nimbus 的作用类似于 Hadoop 中 JobTracker 的角色。

每个工作节点（Work Node）上运行一个被称为 Supervisor 的后台程序。Supervisor 负责监听从 Nimbus 分配给它执行的任务，据此启动或停止执行任务的工作进程。每一个工作进程执行一个 Topology 的子集；一个运行中的 Topology 由分布在不同工作节点上的多个工作进程组成。

Nimbus 和 Supervisor 节点之间所有的协调工作是通过 Zookeeper 集群来实现的，其体系结构如图 10.2 所示。此外，Nimbus 和 Supervisor 进程都是快速失败（fail-fast）和无状态（stateless）的；Storm 集群所有的状态要么在 Zookeeper 集群中，要么存储在本地磁盘上。这意味着你可以用 kill 命令来杀死 Nimbus 和 Supervisor 进程，它们在重启后可以继续工作。这个设计使得 Storm 集群拥有不可思议的稳定性。

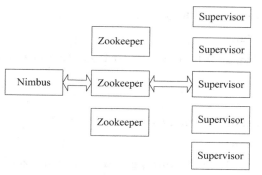

图 10.2　Storm 体系结构

3．启动 Storm 各个后台进程的注意事项

- Storm 后台进程被启动后，将在 Storm 安装部署目录下的 logs/子目录下生成各个进程的日志文件。
- 测试，Storm UI 必须和 Storm Nimbus 部署在同一台机器上，否则 UI 无法正常工作，因为 UI 进程会检查本机是否存在 Nimbus 链接。
- 一般来说，为了方便使用，建议将 storm/bin 加入到系统环境变量中。

10.2.3　任务实施

步骤 1

启动 Zookeeper 服务。执行 zkServer.sh start 启动服务；执行 zkServer.sh stop 停止服务；执行 zkServer.sh status 检查启动情况和模式 leader 和 follower，zookeeper 会随机选择一个节点作为 leader，其他作为 follower。

（1）启动服务。

分别以 root 用户登录 node1、node2、node3、node4，启动 Zookeeper，操作命令如下。

```
[root@node1 ~]# zkServer.sh start
JMX enabled by default
Using config: /home/local/zookeeper/bin/../conf/zoo.cfg
Starting zookeeper ... STARTED
```

其他 3 台节点机都要启动 zookeeper，操作一样。

（2）查看 4 台节点机的服务状态。

```
[root@node1 ~]# zkServer.sh status
JMX enabled by default
Using config: /home/local/zookeeper/bin/../conf/zoo.cfg
Mode: follower


[root@node2 ~]# zkServer.sh status
JMX enabled by default
Using config: /home/local/zookeeper/bin/../conf/zoo.cfg
Mode: follower
```

```
[root@node3 ~]# zkServer.sh status
JMX enabled by default
Using config: /home/local/zookeeper/bin/../conf/zoo.cfg
Mode: leader

[root@node4 ~]# zkServer.sh status
JMX enabled by default
Using config: /home/local/zookeeper/bin/../conf/zoo.cfg
Mode: follower
```

其中节点机 node3 是领导者，谁是领导者是由 zookeeper 分配，领导者只有一个。

步骤 2

启动 Nimbus、Supervisor、UI。在生产环境中主节点机一般不启动 Supervisor 进程。

登录 node1 节点机，启动 Nimbus、Supervisor、UI 进程。操作命令如下。

```
[root@node1 ~]# storm nimbus &
[root@node1 ~]# storm supervisor &
[root@node1 ~]# storm ui &
```

分别登录 node2、node3、node4 节点机，启动 Supervisor 进程。操作命令如下。

```
[root@node2 ~]# storm supervisor &
[root@node3 ~]# storm supervisor &
[root@node4 ~]# storm supervisor &
```

步骤 3

打开浏览器，在地址栏处输入 http://192.168.23.111:8080，查看 Storm 的运行状况，如图 10.3 所示。

图 10.3　web 监控界面

通过该页面可以继续查看到 Nimbus Configuration 的相关信息，如表 10.3 所示。

表 10.3　Nimbus 部分配置信息

Key	Value	说明
storm.local.dir	/home/local/storm/temp	Storm 临时文件存放的地方
nimbus.host	192.168.23.111	nimbus 主机的 IP 地址
storm.zookeeper.root	/home/local/zookeeper	在 root 用户下运行 zookeeper
supervisor.enable	true	从节点服务已启动
storm.zookeeper.servers	["192.168.23.111" "192.168.23.112" "192.168.23.113" "192.168.23.114"]	Storm 系统下各节点所对应的 IP 地址
supervisor.slots.ports	[6700 6701 6702 6703]	JVM 通过这些端口进行交互，一般设置 4 个端口，即在每个节点上可以启动 4 个 JVM
ui.port	8080	网站的端口号
storm.cluster.mode	distributed	采用分布式处理技术

步骤 4

（1）启动 storm 命令，操作命令如下。

```
[root@node1 hbase]# storm
Commands:
        activate
        classpath
        deactivate
        dev-zookeeper
        drpc
        help
        jar
        kill
        list
        localconfvalue
        logviewer
        monitor
        nimbus
        rebalance
        remoteconfvalue
        repl
        shell
        supervisor
        ui
```

```
    version

Help:
    help
    help <command>
```

（2）Storm 的基本命令如下。

```
storm activate topology-name    #激活指定的拓扑 Spout。
storm classpath                 #打印出 Storm 客户端运行命令时使用的类路径。
storm deactivate topology-name  #禁用指定的拓扑 Spout。
storm dev-zookeeper             #以 dev.zookeeper.path 配置的值作为本地目录，
                                #以 storm.zookeeper.port 配置的值作为端口，
                                #启动一个新的 ZooKeeper 服务，仅用来开发/测试。
storm drpc                      #启动一个 DRPC 守护进程。
                                #该命令应该使用 daemontools 或者 monit 工具监控运行。
storm help                      #打印一条帮助消息或者可用命令的列表，语法如下。
storm help <command>            #直接输入不带参数的 storm，也可以启动 storm help 命令。
storm jar topology-jar-path class ...  #运行类的指定参数的 main 方法。
                                #把 Storm 的 jar 文件和"~/.storm"的配置放到类路径中，
                                #以便当拓扑提交时，StormSubmitter 会上传
                                #topology-jar-path 的 jar 文件。
storm kill topology-name [-w wait-time-secs]  #杀死名为 topology-name 的拓扑。
                                #Storm 首先会在拓扑的消息超时时间期间禁用 Spout，
                                #以允许所有正在处理的消息完成处理。然后，Storm 将会
                                #关闭 Worker 并清理它们的状态。可以使用-w 标记覆盖
                                #Storm 在禁用与关闭期间等待的时间长度。
storm list                      #列出正在运行的拓扑及其状态。
storm localconfvalue conf-name  #打印出本地 Storm 配置的 conf-name 的值。
                                #本地 Storm 配置是~/.storm/storm.yaml
                                #与 defaults.yaml 合并的结果。
storm logviewer                 #启动 Logviewer 守护进程。
                                #Logviewer 提供一个 Web 接口查看 Storm 日志文件。
                                #该命令应该使用 daemontools 或者 monit 工具监控运行。
storm nimbus                    #启动 Nimbus 守护进程。
                                #该命令应该使用 daemontools 或者 monit 工具监控运行。
storm rebalance topology-name [-w wait-time-secs]
                                #有时你可能希望扩散一些正在运行的拓扑的 worker。例如，
                                #你有一个 10 个节点的集群，每个节点运行 4 个 worker，
                                #然后假设需要添加另外 10 个节点到集群中。你可能希望有
```

```
                                #Spout 扩散正在运行中的拓扑的 worker，这样每个节点运行
                                #两个 worker。解决的一种方法是杀死拓扑并重新提交拓扑，
                                #但 Storm 提供了一个 rebalance 的命令，我们可以用一种
                                #更简单的方法来做到这一点。rebalance 首先会在消息超时
                                #时间内禁用拓扑，使用-w 可以覆盖超时时间，然后重新均
                                #衡分配集群的 worker，拓扑会返回到它原来的状态，即
                                #禁用的拓扑仍将禁用，激活的拓扑继续激活。
storm remoteconfvalue conf-name    #打印出远程集群 Storm 配置的 conf-name 的值。
                                #集群 Storm 配置是$STORM-PATH/conf/storm.yaml 与
                                #defaults.yaml 合并的结果。该命令必须在集群节点上运行
storm repl                      #打开一个包含类路径中的 jar 文件和配置的 Clojure REPL，
                                #以便调试时使用。
                                #Clojure 可以作为一种脚本语言内嵌到 Java 中，但是 Clojure
                                #的首选编程方式是使用 REPL，REPL 是一个简单的命令行接口。
                                #使用 REPL，可以输入命令并执行，然后查看结果。
storm shell resourcesdir command args    #执行 Shell 脚本。
storm supervisor                #启动 Supervisor 守护进程。
storm ui                        #启动 UI 守护进程。
                                #UI 为 Storm 集群提供了一个 Web 界面并显示运行拓扑的详细
                                #统计信息。该命令应该使用 daemontools 或者 monit 工具监控
                                #运行。
storm version                   #打印 Storm 发布的版本号。
```

10.3 本章小结

通过完成本章任务一和任务二的内容，可以快速掌握大数据流式实时处理系统 Storm 的安装、部署及 UI 管理。本章重点介绍了依赖软件 Python 的编译、安装及使用，项目管理工具 Maven 的安装、配置及使用，版本控制工具 Git 的编译及安装，消息队列服务 ZeroMQ 的编译及安装，JZMQ 的编译及安装，分布式应用程序协调服务 ZooKeeper 的安装、配置及使用。同时，还详细介绍了如何启动、停止 Storm 服务和 Zookeeper 服务，以及使用 Web 对 Storm 进行监控管理。

第 11 章 Storm 拓扑实例

Storm 可以方便地在一个计算机集群中编写与扩展复杂的实时计算,并且能高速地、可靠地处理到来的每个消息。更棒的是你可以使用任意编程语言来做开发。

目前 Storm 已被广泛应用到许多领域,如实时分析、在线机器学习、信息流处理(使用 Storm 处理新的数据和快速更新数据库)、连续性的计算(使用 Storm 连续查询,然后将结果返回给客户端,如将微博上的热门话题转发给用户)、分布式 RPC(远过程调用协议,通过网络从远程计算机程序上请求服务)、ETL(Extraction Transformation Loading,数据抽取、转换和加载)等。

为了让用户尽快掌握 Storm 的使用方法,Storm 的创始人 Nathan Marz 开发了一个让 Storm 用户快速入门的项目——storm-starter,这个项目里有很多适合初学者动手练习的 Topology 示例,如 ExclamationTopology、WordCountTopology、ReachTopology 等,storm-starter 项目详情可登录 GitHub 官网进行查看。

使用 storm-starter 中的 Topology 之前,首先需要安装编译该项目的软件。编译 storm-starter 项目有两种方法,一种是使用 Leiningen,另一种是使用 Maven。Leiningen 是一个用于自动化构建 clojure 项目的工具,而 Maven 是一个基于项目对象模型(POM)的项目管理工具,这两种工具都可以用于项目管理,本章介绍的是第二种方法。

11.1 任务一 完成实例 Storm-Starter

11.1.1 任务描述

本节任务主要完成系统自带的 Storm-Starter 拓扑的运行,检查拓扑运行状态,观察各节点运行信息,使用 mvn 对 Storm-Starter 项目进行打包及运行,并通过完成该任务了解 Storm 的工作流程和熟知 StormUI 管理。

11.1.2 相关知识

开发 Storm 需要了解一些术语:Tuple(元组)、Stream、Spout、Topology、Bolt、Task、Worker、Stream Grouping(消息分组策略)和 Topology(拓扑)。

Tuple:一次消息传递的基本单元。原本是一个 Key-Value 的 map,但是由于各个组件间传递的 tuple 的字段名称已经事先定义好，所以 tuple 中只需按序填入各个 value 即可。

Stream：被处理的数据。通过源源不断传递的 tuple 就组成了 stream。

Spout：在一个 topology 中产生源数据流的组件。通常情况下 Spout 会从外部数据源中读取数据，然后转换为 topology 内部的源数据。Spout 是一个主动的角色，其接口中有个 nextTuple() 函数，Storm 框架会不停地调用此函数，用户只要在其中生成源数据即可。

Topology：在 Storm 中运行的一个实时应用程序，为各个组件间的消息流动形成逻辑上的一个拓扑结构。

Bolt：在一个 topology 中接受数据然后执行处理的组件。Bolt 可以执行过滤、函数操作、合并、写数据库等任何操作。Bolt 是一个被动的角色，其接口中有个 execute（Tuple input）函数，在接受到消息后会调用此函数，用户可以在其中执行自己想要的操作。

Task：运行于 Spout 或 Bolt 中的线程。

Worker：运行这些线程的进程。

Stream Grouping：规定了 Bolt 接收什么样的流作为输入数据，并通过 Stream Grouping 来定义一个 Stream 应该如何分配给 Bolts 上面的多个 Tasks。Storm 里面有 6 种类型的 Stream Grouping。

（1）随机分组（Shuffle Grouping）：随机派发 Stream 里面的 tuple，保证每个 Bolt 接收到的 tuple 数目相同。

（2）字段分组（Fields Grouping）：根据指定字段分割数据流，并分组。比如按 userid 来分组，具有同样 userid 的 tuple 会被分到相同的 Bolts，而不同的 userid 则会被分配到不同的 Bolts。

（3）广播发送（All Grouping）：对于每一个 tuple，所有的 Bolts 都会收到。

（4）全局分组（Global Grouping）：这个 tuple 被分配到 Storm 中一个 Bolt 的其中一个 Task。再具体一点就是分配给 id 值最低的那个 Task。

（5）不分组（None Grouping）：这个分组的意思是 Stream 不关心到底谁会收到它的 tuple。目前这种分组和 Shuffle Grouping 是一样的效果。

（6）直接分组（Direct Grouping）：这是一种比较特别的分组方法，用这种分组意味着消息的发送者指定由消息接收者的哪个 Task 处理这个消息。只有被声明为 Direct Stream 的消息流可以声明这种分组方法，而且这种消息 tuple 必须使用 emitDirect 方法来发送。消息处理者可以通过 TopologyContext 来获取处理它的消息的 taskid。

Topology：由 Stream Grouping 连接起来的 Spout 和 Bolt 节点网络。

11.1.3 任务实施

步骤 1

登录 node1 节点机，进入实例目录，操作如下。

```
[root@node1 ~]# cd /home/local/storm/examples/storm-starter/
[root@node1 storm-starter]# ll
总用量 3204
drwxr-xr-x. 3 root root    4096 11月 19 05:05 multilang
-rw-r--r--. 1 root root    5180 11月 19 11:42 pom.xml
```

```
-rw-r--r--.  1 root root     6526 11月 19 05:05 README.markdown
drwxr-xr-x.  4 root root     4096 11月 19 05:05 src
-rw-r--r--.  1 root root  3248859 11月 19 12:06 storm-starter-topologies-0.9.3.jar
drwxr-xr-x.  3 root root     4096 11月 19 05:05 test
```

步骤 2

运行 storm-starter-topologies 程序，操作如下。

```
[root@node1 storm-starter]# storm jar storm-starter-topologies-0.9.3.jar \
storm.starter.WordCountTopology wordcountTop
```

运行后显示如下信息。

```
File 'storm-starter-topologies-0.9.3.jar' uploaded to '/home/local/storm/
temp/nimbus/inbox/stormjar-87013a7d-41b5-4a6f-b2fd-ede0343994ec.jar' (324
8859 bytes)
1040 [main] INFO backtype.storm.StormSubmitter - Successfully uploaded t
opology jar to assigned location: /home/local/storm/temp/nimbus/inbox/sto
rmjar-87013a7d-41b5-4a6f-b2fd-ede0343994ec.jar
1040 [main] INFO backtype.storm.StormSubmitter - Submitting topology wor
dcountTop in distributed mode with conf {"topology.workers":3,"topology.d
ebug":true}
1215 [main] INFO backtype.storm.StormSubmitter - Finished submitting top
ology: wordcountTop
```

步骤 3

打开浏览器，输入 IP 地址，如图 11.1 所示。

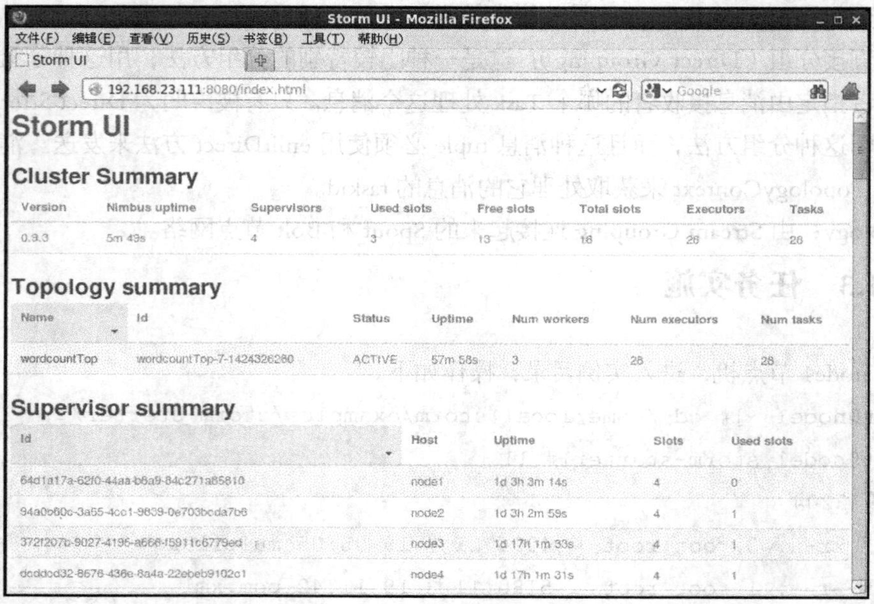

图 11.1 Storm UI

步骤 4

单击 wordcountTop，查看 Topology 信息，如图 11.2 所示。

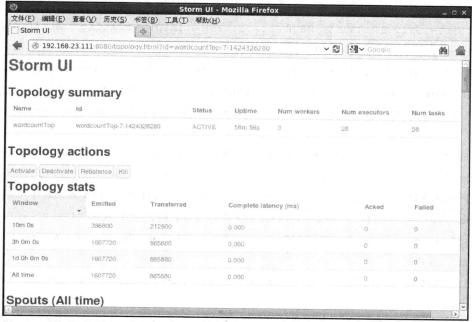

图 11.2 wordcountTop 拓扑信息

步骤 5

节点机运行情况如下，当 Storm 在运行程序时，计算节点机就会自动启动工作进程下载并执行任务。

```
[root@node1 storm]# jps         [root@node2 storm]# jps
16782 Jps                       39969 worker
16633 core                      40126 Jps
43238 QuorumPeerMain            33834 supervisor
53976 supervisor                30293 QuorumPeerMain
16563 nimbus


[root@node3 storm]# jps         [root@node4 storm]# jps
49792 Jps                       49147 worker
49606 worker                    30308 supervisor
30733 QuorumPeerMain            30716 QuorumPeerMain
30231 supervisor                49304 Jps
```

登录 node2 节点机，查看数据日志信息，如下所示。

```
[root@node2 storm]# cd /home/storm/logs
[root@node2 logs]# tail worker-6703.log.1
2015-02-19T15:10:38.964+0800 b.s.d.executor [INFO] Processing received
```

```
message source: spout:25, stream: default, id: {}, [the cow jumped over the moon]
    2015-02-19T15:10:38.965+0800 b.s.d.task [INFO] Emitting: split default
["the"]
    2015-02-19T15:10:38.966+0800 b.s.d.task [INFO] Emitting: split default
["cow"]
    2015-02-19T15:10:38.966+0800 b.s.d.executor [INFO] Processing received
message source: split:19, stream: default, id: {}, ["cow"]
    2015-02-19T15:10:38.966+0800 b.s.d.task [INFO] Emitting: count default [cow,
23985]
    2015-02-19T15:10:38.967+0800 b.s.d.task [INFO] Emitting: split default
["jumped"]
    2015-02-19T15:10:38.967+0800 b.s.d.executor [INFO] Processing received
message source: split:19, stream: default, id: {}, ["jumped"]
    2015-02-19T15:10:38.967+0800 b.s.d.task [INFO] Emitting: count default
[jumped, 23985]
    2015-02-19T15:10:38.967+0800 b.s.d.task [INFO] Emitting: split default
["over"]
    2015-02-19T15:10:38.968+0800 b.s.d.task [INFO] Emitting: split default
["the"]
```

步骤 6

执行 strom list 命令，操作如下。

```
[root@node1storm]# storm list
```

运行结果部分信息显示如下。

```
2050 [main] INFO   backtype.storm.thrift - Connecting to Nimbus at node1:6627
Topology_name       Status     Num_tasks  Num_workers  Uptime_secs
-------------------------------------------------------------------
wordcountTop        ACTIVE     28         3            4269
```

步骤 7

停止 Topology，操作命令如下。

```
[root@node1storm]# storm kill wordcountTop
```

停止 Topology 后，显示的部分信息如下。

```
.6.jar:/home/local/storm/conf:/home/local/storm/bin backtype.storm.command.
kill_topology wordcountTop
    1022 [main] INFO   backtype.storm.thrift - Connecting to Nimbus at node1:6627
    1087 [main] INFO   backtype.storm.command.kill-topology - Killed topology:
wordcountTop
```

步骤 8

Maven 管理 storm-starter 项目，Maven 除了以程序构建能力为特色之外，还提供高级项目管理工具。使用 Maven 将项目打包操作如下。

```
[root@node1 storms]# cd /home/local/storm/examples/storm-starter/
[root@node1 storm-starter]# mvn package
```

打包后显示如下信息，表示打包成功。

```
[INFO] Building jar: /home/local/storm/examples/storm-starter/target/storm-starter-0.9.3-jar-with-dependencies.jar
[INFO] META-INF/ already added, skipping
[INFO] META-INF/MANIFEST.MF already added, skipping
[INFO] twitter4j/ already added, skipping
[INFO] META-INF/LICENSE.txt already added, skipping
[INFO] META-INF/maven/ already added, skipping
[INFO] META-INF/maven/org.twitter4j/ already added, skipping
[INFO] META-INF/ already added, skipping
[INFO] META-INF/MANIFEST.MF already added, skipping
[INFO] META-INF/LICENSE.txt already added, skipping
[INFO] META-INF/maven/ already added, skipping
[INFO] META-INF/MANIFEST.MF already added, skipping
[INFO] META-INF/ already added, skipping
[INFO] META-INF/maven/ already added, skipping
[INFO] META-INF/ already added, skipping
[INFO] META-INF/MANIFEST.MF already added, skipping
[INFO] META-INF/maven/ already added, skipping
[INFO] ------------------------------------------------------------------------
[INFO] BUILD SUCCESS
[INFO] ------------------------------------------------------------------------
[INFO] Total time: 40.315 s
[INFO] Finished at: 2015-02-19T16:25:03+08:00
[INFO] Final Memory: 23M/313M
[INFO] ------------------------------------------------------------------------
```

打包完成后，在 storm-starter 目录下多出一个 target 目录，并且在 target 目录中生成两个 jar 文件，如下所示。

```
[root@node1 storm-starter]# cd /home/local/storm/examples/storm-starter/target
[root@node1 target]# ll
总用量 3360
drwxr-xr-x. 2 root root   4096 2月  19 16:25 archive-tmp
drwxr-xr-x. 6 root root   4096 2月  19 16:07 classes
drwxr-xr-x. 3 root root   4096 2月  19 16:07 generated-sources
```

```
drwxr-xr-x. 3 root root     4096 2月  19 16:07 generated-test-sources
drwxr-xr-x. 2 root root     4096 2月  19 16:08 maven-archiver
drwxr-xr-x. 3 root root     4096 2月  19 16:06 maven-shared-archive-resources
drwxr-xr-x. 3 root root     4096 2月  19 16:07 maven-status
-rw-r--r--. 1 root root   143698 2月  19 16:24 storm-starter-0.9.3.jar
-rw-r--r--. 1 root root  3248731 2月  19 16:25 storm-starter-0.9.3-jar-with-
dependencies.jar
drwxr-xr-x. 2 root root     4096 2月  19 16:25 surefire
drwxr-xr-x. 5 root root     4096 2月  19 16:08 surefire-reports
drwxr-xr-x. 4 root root     4096 2月  19 16:07 test-classes
```

步骤9

提交运行，操作如下。

```
[root@node1 target]# storm jar storm-starter-0.9.3-jar-with-dependencies.jar storm.starter.WordCountTopology wcTop
```

运行结果如下。

```
File 'storm-starter-0.9.3-jar-with-dependencies.jar' uploaded to '/home/local/
storm/temp/nimbus/inbox/stormjar-2f794f2b-7b73-4248-b3e6-e9ecebea2bde.jar' (3248731 bytes)
886 [main] INFO  backtype.storm.StormSubmitter - Successfully uploaded topology jar to assigned location:
/home/local/storm/temp/nimbus/inbox/ stormjar-2f794f2b-7b73-4248-b3e6-e9ecebea2bde. jar
887 [main] INFO  backtype.storm.StormSubmitter - Submitting topology wcTop in distributed mode with conf {"topology.workers":3,"topology.debug":true}
1199 [main] INFO  backtype.storm.StormSubmitter - Finished submitting topology: wcTop
```

步骤10

停止 wcTop 运行，操作如下。

```
[root@node1 target]# storm kill wcTop
```

运行结果如下。

```
.6.jar:/home/local/storm/conf:/home/local/storm/bin backtype.storm.command.kill_topology wcTop
1023 [main] INFO  backtype.storm.thrift - Connecting to Nimbus at node1:6627
1072 [main] INFO  backtype.storm.command.kill-topology - Killed topology: wcTop
```

11.2 任务二 使用 Eclipse 管理 Storm-Starter

11.2.1 任务描述

本节任务主要是使用 Eclipse 软件来管理 Storm-Starter 项目，熟悉 Eclipse 软件的编程环境，通过完成 Storm-Starter 项目进一步加深了解 Storm 拓扑的编程方法及 Web 监控管理等。

11.2.2 相关知识

Eclipse 是一个开放源代码的、基于 Java 的可扩展开发平台，于 1999 年 4 月由 OTI 和 IBM 两家公司的 IDE 产品开发组创建，是替代商业软件 Visual Age for Java 的下一代 IDE 开发环境。2001 年 11 月贡献给开源社区，现在它由非营利软件供应商联盟 Eclipse 基金会（Eclipse Foundation）管理。

Eclipse 是著名的跨平台的自由集成开发环境（IDE）。最初主要用于 Java 语言开发。由于其源码开放，任何人都可以免费获得，而且在此基础上还可开发各自的插件，并通过安装不同的插件 Eclipse 可以支持不同的计算机语言，比如 C++ 和 Python 等开发工具。

Eclipse 的本身只是一个框架平台，但是众多插件的支持使得 Eclipse 拥有其他功能相对固定的 IDE 软件很难具有的灵活性。许多软件开发商是以 Eclipse 为框架开发自己的 IDE，随后许多大公司（包括 Oracle）也纷纷加入了该项目。目前，Eclipse 的目标是成为可进行任何语言开发的 IDE 集成者，使用者只需下载各种语言的插件即可，因此，越来越受到人们的关注。

熟悉 Eclipse 快捷键的操作可以使开发工作事半功倍，节省更多的时间来做有意义的事情。常用的快捷键操作如下所示。

F11：调试程序。

Ctrl+F11：运行程序。

Ctrl+1：快速修复。

Ctrl+D：删除当前行。

Alt+↓：当前行和下面一行交换位置。

Alt+↑：当前行和上面一行交换位置。

Alt+←：切换编辑窗口到前一个编辑页面。

Alt+→：切换编辑窗口到下一个编辑页面。

Alt+Enter：显示当前选择资源的属性。

Shift+Enter：在当前行的下一行插入空行。

Shift+Ctrl+Enter：在当前行插入空行。

Ctrl+Q：定位到最后编辑的地方。

Ctrl+L：定位在某行。

Ctrl+M：最大化当前的 Edit 或 View（再按则还原）。

Ctrl+/：注释当前行（再按则取消注释）。

Ctrl+O：快速显示 OutLine。

Ctrl+T：快速显示当前类的继承结构。

Ctrl+W：关闭当前 Editer。

Ctrl+K：参照选中的 Word 快速定位到下一个。

Ctrl+E：快速显示当前 Editer 的下拉列表。

Ctrl+(小键盘的/)：折叠或展开当前类中的所有代码。

Alt+/：代码助手完成一些代码的插入。

Shift+Ctrl+E：显示管理当前打开的所有的 View 的管理器（可以选择关闭，激活等操作）。

Ctrl+J：正向增量查找（按 Ctrl+J 后，你所输入的每个字母编辑器都提供快速匹配定位到某个单词，如果没有，则在 stutes line 中显示没有找到。查找单词时，特别实用）。

Shift+Ctrl+J：反向增量查找（和上条相同，只不过是从后往前查）。

Shift+Ctrl+S：保存所有打开的 Editer。

Shift+Ctrl+F4：关闭所有打开的 Editer。

Shift+Ctrl+X：把当前选中的文本全部变为大写。

Shift+Ctrl+Y：把当前选中的文本全部变为小写。

Shift+Ctrl+F：程序代码自动排版。

Shift+Ctrl+P：定位到对于的匹配符（譬如{}）。

11.2.3 任务实施

步骤 1

将 storm-starter 编译成 eclipse 工程，操作如下。

```
[root@node1 ~]# cd /home/local/storm/examples/storm-starter/
[root@node1 storm-starter]# mvn eclipse:eclipse
```

编译后结果如下。

```
[INFO] Using Eclipse Workspace: null
[INFO] Adding default classpath container: org.eclipse.jdt.launching.JRE_CONTAINER
[INFO] Wrote settings to /home/local/storm/examples/storm-starter/.settings/org.eclipse.jdt.core.prefs
[INFO] Wrote Eclipse project for "storm-starter" to /home/local/storm/examples/storm-starter.
[INFO]
[INFO] ------------------------------------------------------------
[INFO] BUILD SUCCESS
[INFO] ------------------------------------------------------------
[INFO] Total time: 02:02 min
[INFO] Finished at: 2015-02-19T16:44:50+08:00
[INFO] Final Memory: 22M/312M
[INFO] ------------------------------------------------------------
```

步骤 2

运行 Eclipse，操作如下。

```
[root@node1 ~]# /usr/local/eclipse/eclipse &
```

输入工作目录，如图 11.3 所示。

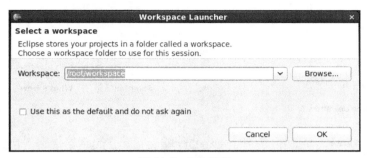

图 11.3 工作目录

步骤 3

单击"OK"后，进入工作窗口，如图 11.4 所示。

图 11.4 eclipse 工作窗口

步骤 4

打开 Eclipse 后，单击"File—Import..."，如图 11.5 所示。

步骤 5

打开 Import 窗口后，单击" General"—"Existing Projects into WorkSpace"，如图 11.6 所示。

步骤 6

单击"Next>"后，选择"/home/local/storm/examples/storm-starter"项目，如图 11.7 所示。

图 11.5　eclipse 项目导入

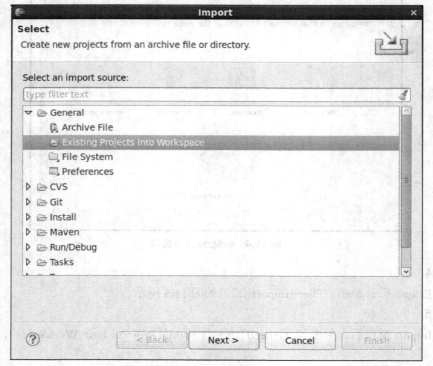

图 11.6　eclipse 选择项目

步骤 7

单击"Finish"后,打开项目窗口,如图 11.8 所示。

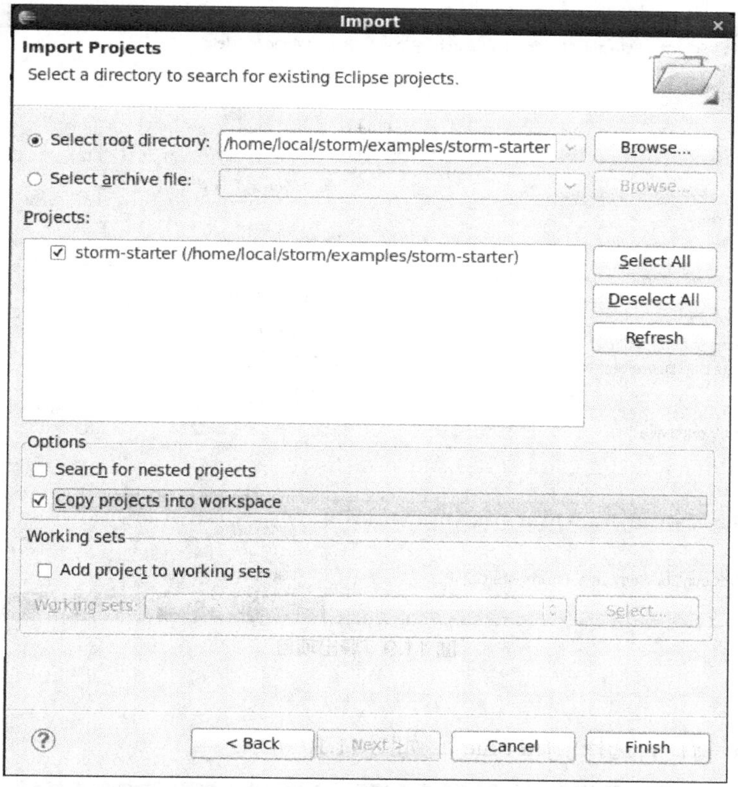

图 11.7 导入项目

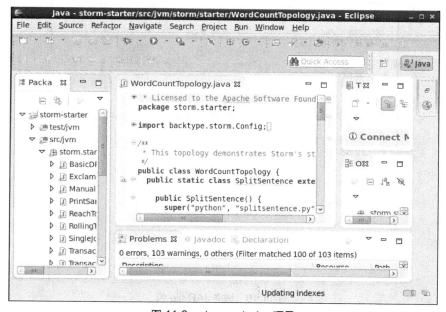

图 11.8 storm-starter 项目

步骤 8

单击"File"—"Export...",如图 11.9 所示。

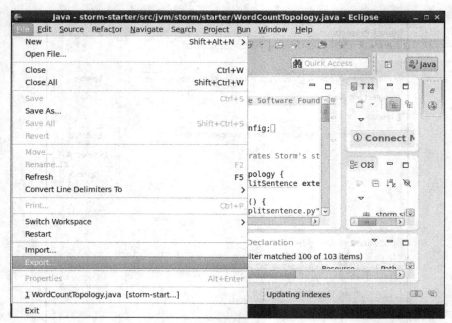

图 11.9 导出项目

步骤 9

打开 Export 窗口，选择 "JAR file"，如图 11.10 所示。

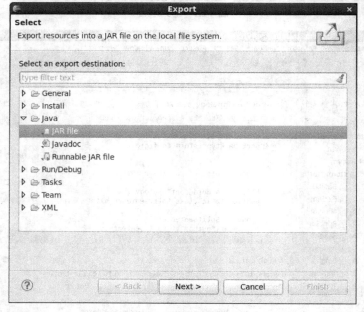

图 11.10 选择 JAR file 包

步骤 10

单击 "Next>" 后，选择 storm-starter 项目中的 src/jvm、multilang、src 全部包，输入 JAR file 文件名，如图 11.11 所示。

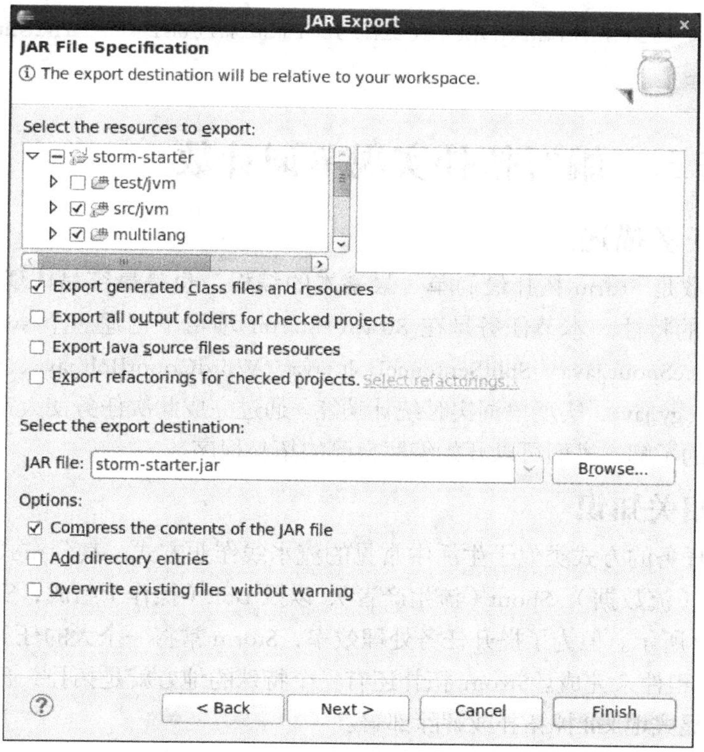

图 11.11　导出 storm-starter 包

单击"Finish",生成 jar 包。如下所示。

```
[root@node1 ~]# cd /root/workspace/
[root@node1 workspace]# ll
总用量 88
drwxr-xr-x. 7 root root  4096 2月  19 17:06 storm-starter
-rw-r--r--. 1 root root 82312 2月  19 18:15 storm-starter.jar
```

步骤 11

运行 storm-wordcount.jar,操作命令如下。

```
[root@node1 workspace]# storm jar storm-starter.jar storm.starter   \
.WordCountTopology wcTop
```

运行结果如下。

```
File 'storm-starter.jar' uploaded to '/home/local/storm
/temp/nimbus/inbox/stormjar-014728ea-c014-4a43-905a-93dcd036ee2f.jar'
(95912 bytes)
    861 [main] INFO  backtype.storm.StormSubmitter - Successfully uploaded
topology jar to assigned location: /home/local/storm/temp/nimbus/inbox/
stormjar-014728ea-c014-4a43-905a-93dcd036ee2f.jar
    861 [main] INFO  backtype.storm.StormSubmitter - Submitting topology wcTop
in distributed mode with conf {"topology.workers":3,"topology.debug":true}
```

996 [main] INFO backtype.storm.StormSubmitter - Finished submitting topology: wcTop

11.3 任务三 编写拓扑实现单词计数

11.3.1 任务描述

实现单词计数是 Storm 拓扑最简单、最基本的编程,也是最容易体现出 Storm 对数据流进行实时处理的特性。本节任务是在 Storm-Starter 项目下创建一个 swpt 包,并编写 5 个类文件 SentenceSpout.java、SplitSentenceBolt.java、WordCountBolt.java、ReportBolt.java、WordCountTopology.java,实现单词实时统计功能。通过完成此次任务使大家对 Storm 的运行过程能有更深入的了解,进而可以开始编写自己的拓扑程序。

11.3.2 相关知识

Storm 处理任务的方式类似于生活中常见的流水线作业方式。其分布式计算结构称为拓扑,它由 Stream(流数据)、Spout(流生产者)、以及 Bolt(操作)组成,实现一个任务的完整拓扑如图 11.12 所示。但为了提升任务处理效率,Storm 常将一个大的任务拆分为几部分,并由不同的 Bolt 组件来完成。Strom 拓扑还有一个特殊的地方就是拓扑一旦开始运行就不会自动停止,除非显式地 kill 掉拓扑或解除部署。

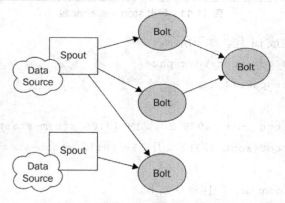

图 11.12 Storm 编程模型拓扑

在本任务中为了方便初学者的学习,所选择的数据来源是一个不变的句子列表,对这些句子遍历,发射出每个句子的元组,然后完成单词统计任务(WordCount),该任务的拓扑如图 11.13 所示。然而,在真实的应用环境中,一个 Spout 通常连接到一个动态数据源,如从 Twitter API 查询到的微博等。下面对这些常用组件做下说明。

Spout 组件:负责读取要统计的数据源中的句子。

Split 组件:负责将接收到的句子拆分成单个的单词,把这些单词发送至 count 组件。

WordCount Bolt:持续对它收到的特定单词计数。每当它收到元组,它将增加与单词相关联计数器,并发出当前这个词和当前记录数。

Report Bolt:该报告 bolt 订阅 WordCount Bolt 类的输出并维护一个表包含所有单词和相应的数量,就像 WordCountBolt 一样。当它收到一个元组,更新表并将内容打印到控制台。

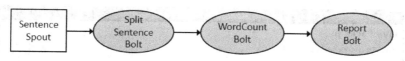

图 11.13 WordCount 拓扑

这样一个统计单词的任务就被拆分为 4 个部分来操作，前 3 个部分可以根据任务的繁重程度来规划并行数目；各组件的并行数没有明确规定，如可以将 Spout 并行数设置为 2，Split 为 4，Count 为 8。本任务中对各组件设置的并行数为默认值。最后一部分是 Report Bolt 类，它存在的主要目的是用于产生每个单词的报告。它与 WordCount Bolt 类一样，使用一个 HashMap 对象来记录数量，但在这种情况下，它只是存储收到 Counter Bolt 的数字。到目前为止，Report Bolt 与其他 Bolt 之间的一个区别是，它是一个终止 Bolt，只接收元组，不会发出任何流，所以也就无并行数设置的必要。

11.3.3 任务实施

步骤 1

右击"storm-starter"—"src/jvm"，打开对话框，单击"New"—"Package"，如图 11.14 所示。

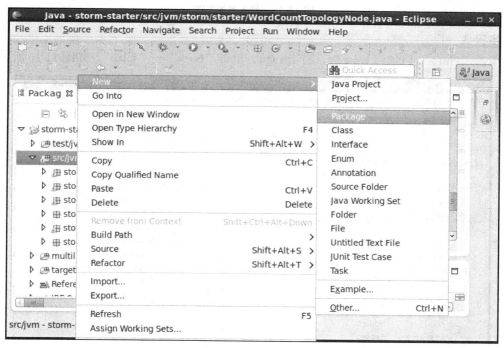

图 11.14 新建包

步骤 2

单击"New"—"Package"后打开对话框，输入 Java Package 的名称，如图 11.15 所示。

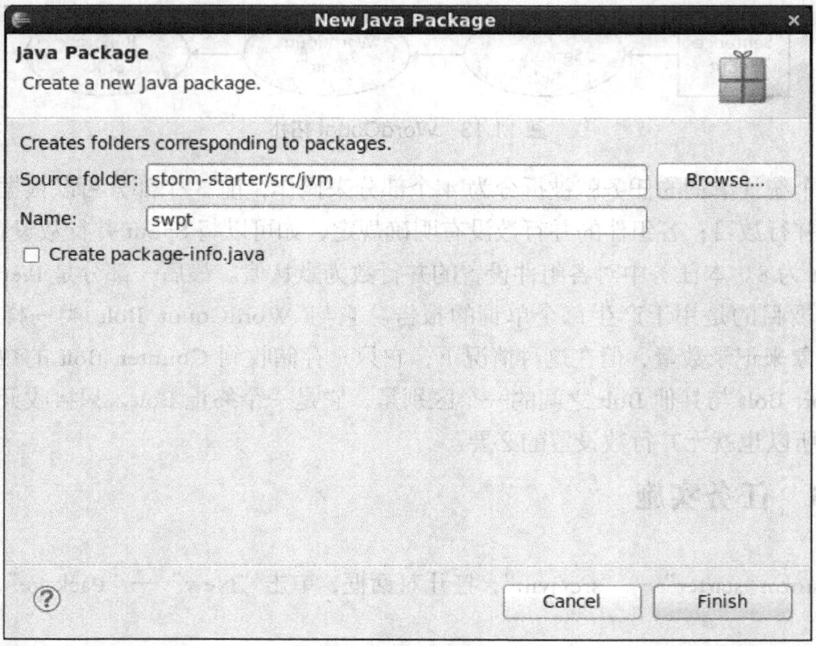

图 11.15 输入包的名称

步骤 3

单击 "Finish" 后创建 swpt 包，右击 swpt，单击 "New" — "Class"，如图 11.16 所示。

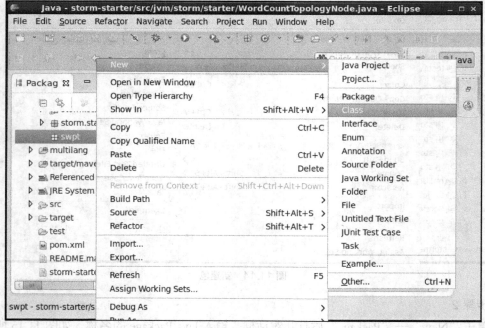

图 11.16 新建类

步骤 4

单击 "New" — "Class" 后，打开对话框，输入 Java Class 的名称，如图 11.17 所示。

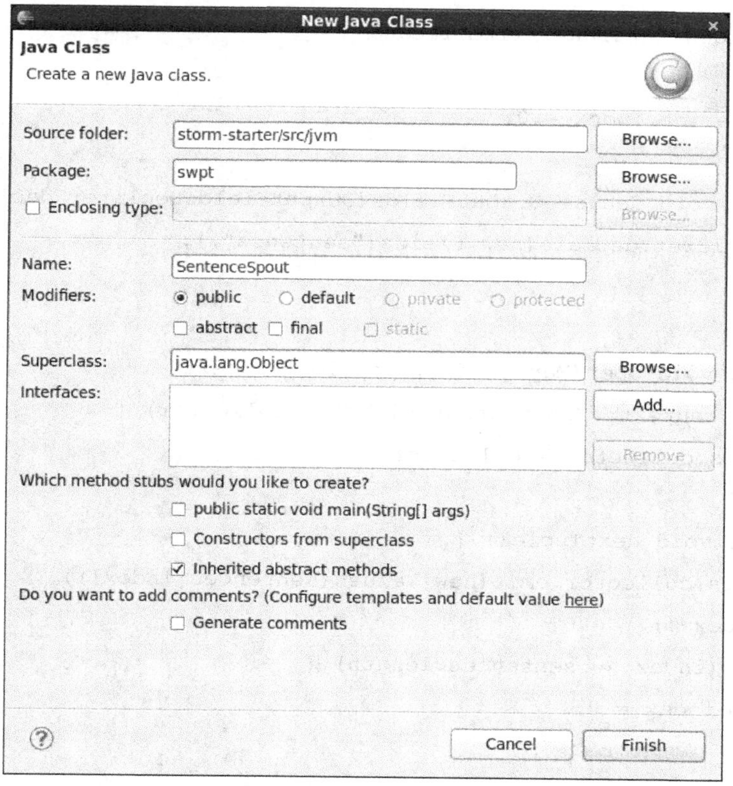

图 11.17 输入类名

单击"Finish"后，在新建的类文件 SentenceSpout.java 中输入如下代码。

```java
package swpt;
import java.util.Map;
import backtype.storm.spout.SpoutOutputCollector;
import backtype.storm.task.TopologyContext;
import backtype.storm.topology.OutputFieldsDeclarer;
import backtype.storm.topology.base.BaseRichSpout;
import backtype.storm.tuple.Fields;
import backtype.storm.tuple.Values;
import backtype.storm.utils.Utils;

public class SentenceSpout extends BaseRichSpout {
    private SpoutOutputCollector collector;
    private String[] sentences = {
        "my dog has fleas",
        "i like swvtc.cn jie",
        "the dog ate my homework",
        "don't have a swvtc.cn jie",
```

```
        "i don't think i like fleas"
    };
    private int index = 0;

    public void declareOutputFields(OutputFieldsDeclarer declarer) {
        declarer.declare(new Fields("sentence"));
    }

    public void open(Map config, TopologyContext
            context, SpoutOutputCollector collector) {
        this.collector = collector;
    }
    public void nextTuple() {
        this.collector.emit(new Values(sentences[index]));
        index++;
        if (index >= sentences.length) {
            index = 0;
        }
        Utils.sleep(1);
    }
}
```

步骤 5
重复步骤 4 创建以下 4 个类文件。

（1）创建类 SplitSentenceBolt.java，代码如下。

```
package swpt;
import java.util.Map;
import backtype.storm.task.OutputCollector;
import backtype.storm.task.TopologyContext;
import backtype.storm.topology.OutputFieldsDeclarer;
import backtype.storm.topology.base.BaseRichBolt;
import backtype.storm.tuple.Fields;
import backtype.storm.tuple.Tuple;
import backtype.storm.tuple.Values;

public class SplitSentenceBolt extends BaseRichBolt {
    private OutputCollector collector;
    public void prepare(Map config, TopologyContext
            context, OutputCollector collector) {
```

```java
        this.collector = collector;
    }

    public void execute(Tuple tuple) {
        String sentence = tuple.getStringByField("sentence");
        String[] words = sentence.split(" ");
        for(String word : words){
            this.collector.emit(new Values(word));
        }
    }
    public void declareOutputFields(OutputFieldsDeclarer declarer) {
        declarer.declare(new Fields("word"));
    }
}
```

（2）创建类 WordCountBolt.java，代码如下。

```java
package swpt;
import java.util.HashMap;
import java.util.Map;
import backtype.storm.task.OutputCollector;
import backtype.storm.task.TopologyContext;
import backtype.storm.topology.OutputFieldsDeclarer;
import backtype.storm.topology.base.BaseRichBolt;
import backtype.storm.tuple.Fields;
import backtype.storm.tuple.Tuple;
import backtype.storm.tuple.Values;

public class WordCountBolt extends BaseRichBolt {
    private OutputCollector collector;
    private HashMap<String, Long> counts = null;
    public void prepare(Map config, TopologyContext
            context, OutputCollector collector) {
        this.collector = collector;
        this.counts = new HashMap<String, Long>();
    }

    public void execute(Tuple tuple) {
        String word = tuple.getStringByField("word");
        Long count = this.counts.get(word);
```

```
        if(count == null){
            count = 0L;
        }
        count++;
        this.counts.put(word, count);
        this.collector.emit(new Values(word, count));
    }

    public void declareOutputFields(OutputFieldsDeclarer declarer) {
        declarer.declare(new Fields("word", "count"));
    }
}
```

（3）创建类 ReportBolt.java，代码如下。

```
package swpt;
import java.util.Collections;
import java.util.HashMap;
import java.util.Map;
import java.util.List;
import backtype.storm.task.OutputCollector;
import backtype.storm.task.TopologyContext;
import backtype.storm.topology.OutputFieldsDeclarer;
import backtype.storm.topology.base.BaseRichBolt;
import backtype.storm.tuple.Tuple;

public class ReportBolt extends BaseRichBolt {
    private HashMap<String, Long> counts = null;

    public void prepare(Map config, TopologyContext context, OutputCollector collector) {
        this.counts = new HashMap<String, Long>();
    }
    public void execute(Tuple tuple) {
        String word = tuple.getStringByField("word");
        Long count = tuple.getLongByField("count");
        this.counts.put(word, count);
    }

    public void declareOutputFields(OutputFieldsDeclarer declarer) {
```

```java
        // this bolt does not emit anything
    }
    public void cleanup() {
        System.out.println("--- FINAL COUNTS ---");
        List<String> keys = new java.util.ArrayList<String>();
        keys.addAll(this.counts.keySet());
        Collections.sort(keys);
        for (String key : keys) {
            System.out.println(key + " : " + this.counts.get(key));
        }
        System.out.println("--------------");
    }
}
```

（4）创建类 WordCountTopology.java，代码如下。

```java
package swpt;
import backtype.storm.Config;
import backtype.storm.LocalCluster;
import backtype.storm.topology.TopologyBuilder;
import backtype.storm.tuple.Fields;
import backtype.storm.utils.Utils;

public class WordCountTopology {

    private static final String SENTENCE_SPOUT_ID = "sentence-spout";
    private static final String SPLIT_BOLT_ID = "split-bolt";
    private static final String COUNT_BOLT_ID = "count-bolt";
    private static final String REPORT_BOLT_ID = "report-bolt";
    private static final String TOPOLOGY_NAME = "word-count-topology";

    public static void main(String[] args) throws
        Exception {
        SentenceSpout spout = new SentenceSpout();
        SplitSentenceBolt splitBolt = new SplitSentenceBolt();
        WordCountBolt countBolt = new WordCountBolt();
        ReportBolt reportBolt = new ReportBolt();

        TopologyBuilder builder = new TopologyBuilder();
        builder.setSpout(SENTENCE_SPOUT_ID, spout);
```

```
        // SentenceSpout --> SplitSentenceBolt
        builder.setBolt(SPLIT_BOLT_ID,
splitBolt).shuffleGrouping(SENTENCE_SPOUT_ID);
        // SplitSentenceBolt --> WordCountBolt
        builder.setBolt(COUNT_BOLT_ID, countBolt).fieldsGrouping(
            SPLIT_BOLT_ID, new Fields("word"));
        // WordCountBolt --> ReportBolt
        builder.setBolt(REPORT_BOLT_ID,
reportBolt).globalGrouping(COUNT_BOLT_ID);
        Config config = new Config();
        LocalCluster cluster = new LocalCluster();
        cluster.submitTopology(TOPOLOGY_NAME, config,
            builder.createTopology());
        Utils.sleep(50000);
        cluster.killTopology(TOPOLOGY_NAME);
        cluster.shutdown();
    }
}
```

步骤 6

右击 "swpt"，单击 "Export..."，如图 11.18 所示。

图 11.18 导出包

步骤 7

单击 "Export..." 后打开对话框，选择 "Java" — "JVM file"，单击 "Next>" 打开 jar 包对话框，选择 "swpt"，选择输入 JAR file 文件名，如图 11.19 所示。

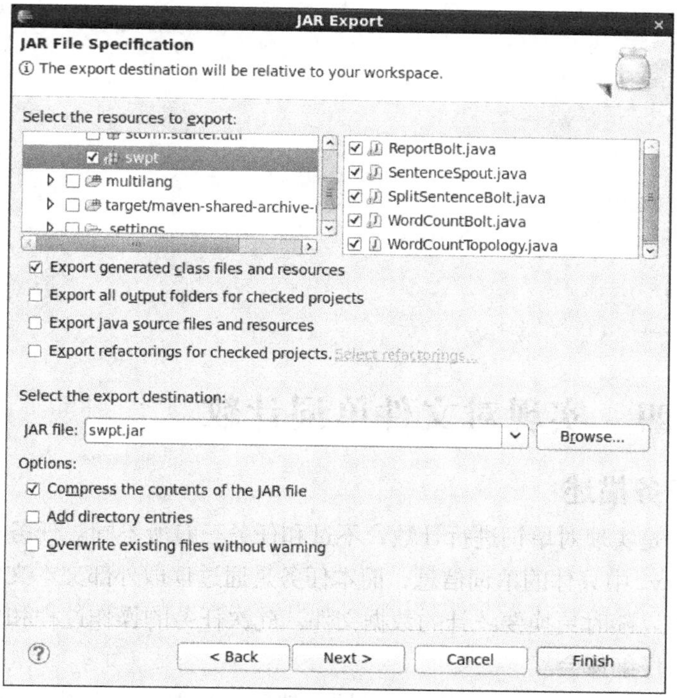

图 11.19　导出 swpt 包

单击"Finish"后，生成如下文件。

```
[root@node1 ~]# cd /root/workspace/
[root@node1 workspace]# ll
总用量 108
drwxr-xr-x. 7 root root  4096 2月  19 17:06 storm-starter
-rw-r--r--. 1 root root 95912 2月  19 18:28 storm-starter.jar
-rw-r--r--. 1 root root  6263 2月  20 15:19 swpt.jar
```

步骤 8

运行测试，操作命令如下。

```
[root@node1 workspace]# storm jar swpt.jar swpt.WordCountTopology wc
```

运行后，在控制台显示部分信息如下。

```
--- FINAL COUNTS ---
a : 6310
ate : 6310
dog : 12620
don't : 12619
fleas : 12619
has : 6310
have : 6310
homework : 6310
```

```
i : 18928
jie : 12620
like : 12619
my : 12620
swvtc.cn : 12620
the : 6310
think : 6309
--------------
```

11.4 任务四 实现对文件单词计数

11.4.1 任务描述

本节任务仍然是实现对单词进行计数,不过和任务三有所不同。任务三是通过读取类文件 SentenceSpout.java 中具体的单词信息,而本任务是通过读取外部文本文件 word.txt 来实现单词统计,这样可以随时更换要统计的数据文件。此次任务的操作过程和任务三类似,但又有所不同,请读者仔细体会。

完成该任务首先需要完成 pom 文件的配置,然后,使用 mvn 生成一个 shanwei-starter 工程项目,进而开始使用 Eclipse 软件来编写相应的类文件 WordReader.java、WordNormalizer.java、WordCounter.java 和 WordCountTop.java。

11.4.2 任务实施

步骤 1

创建一个工程 shanwei-starter,操作如下。

(1) 在 wordspace 下创建相应目录。

```
[root@node1 ~]# mkdir -p ~/workspace/shanwei-starter/src/jvm
[root@node1 ~]# mkdir -p ~/workspace/shanwei-starter/multilang/resources
```

(2) 在目录 shanwei-starter 下编写一个含有基本组件的 pom.xml 文件,内容如下。

```
<?xml version="1.0" encoding="UTF-8"?>
<project xmlns="http://maven.apache.org/POM/4.0.0"
xmlns:xsi="http://www.w3.org/2001/XMLSchema-instance"
xsi:schemaLocation="http://maven.apache.org/POM/4.0.0
http://maven.apache.org/xsd/maven-4.0.0.xsd">
   <modelVersion>4.0.0</modelVersion>

   <groupId>storm.shanwei</groupId>
   <artifactId>shanwei-starter</artifactId>
   <packaging>jar</packaging>
   <version>0.0.1</version>
```

```xml
<name>shanwei-starter</name>

<dependencies>
<dependency>
<groupId>org.apache.storm</groupId>
<artifactId>storm-core</artifactId>
<version>0.9.3</version>
<!-- keep storm out of the jar-with-dependencies -->
<scope>provided</scope>
</dependency>
<dependency>
<groupId>commons-collections</groupId>
<artifactId>commons-collections</artifactId>
<version>3.2.1</version>
</dependency>
</dependencies>

<build>
<sourceDirectory>src/jvm</sourceDirectory>
<resources>
<resource>
<directory>${basedir}/multilang</directory>
</resource>
</resources>

<plugins>
<plugin>
<artifactId>maven-assembly-plugin</artifactId>
<configuration>
<descriptorRefs>
<descriptorRef>jar-with-dependencies</descriptorRef>
</descriptorRefs>
<archive>
<manifest>
<mainClass />
</manifest>
</archive>
</configuration>
```

```xml
<executions>
<execution>
<id>make-assembly</id>
<phase>package</phase>
<goals>
<goal>single</goal>
</goals>
</execution>
</executions>
</plugin>

<plugin>
<groupId>org.codehaus.mojo</groupId>
<artifactId>exec-maven-plugin</artifactId>
<version>1.2.1</version>
<executions>
<execution>
<goals>
<goal>exec</goal>
</goals>
</execution>
</executions>
<configuration>
<executable>java</executable>
<includeProjectDependencies>true</includeProjectDependencies>
<includePluginDependencies>false</includePluginDependencies>
<classpathScope>compile</classpathScope>
<mainClass>${storm.topology}</mainClass>
</configuration>
</plugin>
</plugins>
</build>
</project>
```

步骤2

执行 mvn eclipse:eclipse 命令生成 eclipse 项目文件，如下所示。

```
[root@node1 shanwei-starter]# mvn eclipse:eclipse
[INFO] Wrote settings to /root/workspace/shanwei-starter/.settings/org.eclipse.jdt.core.prefs
```

```
[INFO] Wrote Eclipse project for "shanwei-starter" to /root/workspace
/shanwei-starter.
[INFO]
[INFO] ------------------------------------------------------------
[INFO] BUILD SUCCESS
[INFO] ------------------------------------------------------------
[INFO] Total time: 02:09 min
[INFO] Finished at: 2015-02-22T20:31:10+08:00
[INFO] Final Memory: 22M/312M
[INFO]
------------------------------------------------------------
```

步骤 3

导入项目。打开 eclipse，单击"File"—"Import…"，选择"General"—"Existing Projects into Workspace"，单击"Next>"，选择项目路径，如图 11.20 所示。

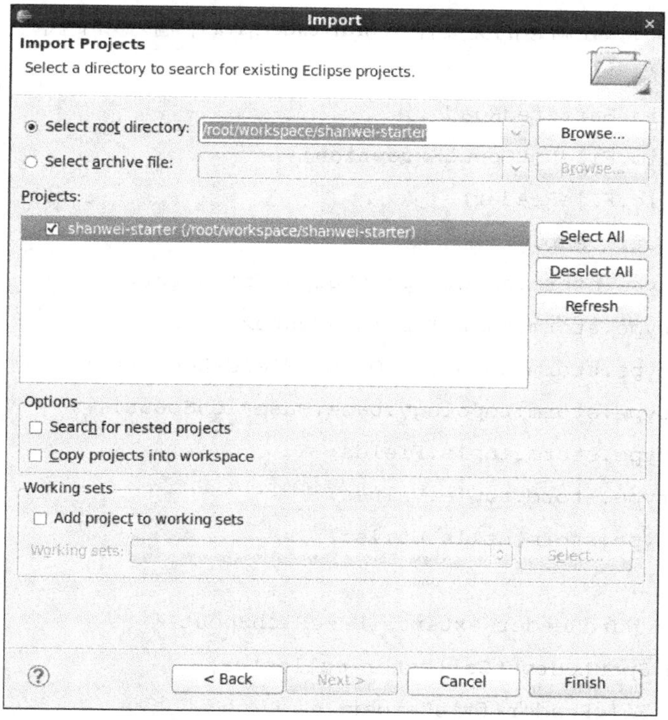

图 11.20　新建 shanwei-starter 项目

步骤 4

创建包和类。右击"storm-starter"—"src/jvm"，打开对话框，单击"New"—"Package"，对话框，输入 Java Package 的名称，如图 11.21 所示。

步骤 5

右击 swpt，单击"New"—"Class"，输入 Java Class 的名称，如图 11.22 所示。

图 11.21　新建 swpt 包　　　　　　图 11.22　新建 WordReader 类

单击"Finish"后，在新建的类文件 WordReader.java 中输入如下代码。

```java
package swpt;
import java.io.BufferedReader;
import java.io.FileNotFoundException;
import java.io.FileReader;
import java.util.Map;
import backtype.storm.spout.SpoutOutputCollector;
import backtype.storm.task.TopologyContext;
import backtype.storm.topology.OutputFieldsDeclarer;
import backtype.storm.topology.base.BaseRichSpout;
import backtype.storm.tuple.Fields;
import backtype.storm.tuple.Values;
import backtype.storm.utils.Utils;

public class WordReader extends BaseRichSpout {
    private SpoutOutputCollector collector;
    private FileReader fileReader;
    private boolean completed = false;
    public void ack(Object msgId) {
        System.out.println("OK:" + msgId);
    }
    public void close() {
    }
```

```java
    public void fail(Object msgId) {
        System.out.println("FAIL:" + msgId);
    }
    public void declareOutputFields(OutputFieldsDeclarer declarer) {
        declarer.declare(new Fields("line"));
    }
    public void nextTuple() {
        if (completed) {
            try {
                Thread.sleep(1000);
            } catch (InterruptedException e) {
            }
            return;
        }
        String str;
        BufferedReader reader = new BufferedReader(fileReader);
        try {
            while ((str = reader.readLine()) != null) {
                this.collector.emit(new Values(str));
            }
        } catch (Exception e) {
            throw new RuntimeException("Error reading tuple", e);
        } finally {
            completed = true;
        }
    }
    public void open(Map conf, TopologyContext context,
            SpoutOutputCollector collector) {
        try {
            this.fileReader = new FileReader(conf.get("wordFile").toString());
        } catch (FileNotFoundException e) {
            throw new RuntimeException("Error reading file ["
                    + conf.get("wordFile") + "]");
        }
        this.collector = collector;
    }
}
```

步骤 6

按照步骤 5，创建其他类。

（1）创建 WordNormalizer.java 类，代码如下。

```java
package swpt;
import java.util.Map;
import backtype.storm.task.OutputCollector;
import backtype.storm.task.TopologyContext;
import backtype.storm.topology.BasicOutputCollector;
import backtype.storm.topology.OutputFieldsDeclarer;
import backtype.storm.topology.base.BaseBasicBolt;
import backtype.storm.topology.base.BaseRichBolt;
import backtype.storm.tuple.Fields;
import backtype.storm.tuple.Tuple;
import backtype.storm.tuple.Values;

public class WordNormalizer extends BaseRichBolt {
    private OutputCollector collector;
    public void prepare(Map config, TopologyContext context,
            OutputCollector collector) {
        this.collector = collector;
    }
    public void execute(Tuple tuple) {
        String sentence = tuple.getStringByField("line");
        String[] words = sentence.split(" ");
        for (String word : words) {
            this.collector.emit(new Values(word));
        }
    }
    public void declareOutputFields(OutputFieldsDeclarer declarer) {
        declarer.declare(new Fields("word"));
    }
}
```

（2）创建 WordCounter.java 类，代码如下。

```java
package swpt;
import java.util.Collections;
import java.util.HashMap;
import java.util.Map;
import java.util.List;
```

```java
import backtype.storm.task.OutputCollector;
import backtype.storm.task.TopologyContext;
import backtype.storm.topology.BasicOutputCollector;
import backtype.storm.topology.OutputFieldsDeclarer;
import backtype.storm.topology.base.BaseBasicBolt;
import backtype.storm.topology.base.BaseRichBolt;
import backtype.storm.tuple.Fields;
import backtype.storm.tuple.Tuple;
import backtype.storm.tuple.Values;
public class WordCounter extends BaseRichBolt {
    private OutputCollector collector;
    Integer id;
    String name;
    Map<String, Long> counters;
    public void prepare(Map config, TopologyContext context,
            OutputCollector collector) {
        this.collector = collector;
        this.counters = new HashMap<String, Long>();
        this.name = context.getThisComponentId();
        this.id = context.getThisTaskId();
    }
    public void execute(Tuple tuple) {
        String word = tuple.getStringByField("word");
        Long count = this.counters.get(word);
        if (count == null) {
            count = 0L;
        }
        count++;
        this.counters.put(word, count);
    }
    public void declareOutputFields(OutputFieldsDeclarer declarer) {
        // this bolt does not emit anything
    }
    public void cleanup() {
        System.out.println("-- Word Counter [" + name + "-" + id + "] --");
        for (Map.Entry<String, Long> entry : counters.entrySet()) {
            System.out.println(entry.getKey() + ": " + entry.getValue());
        }
```

       }
}

（3）创建 WordCountTop.java 类，代码如下。

```java
package swpt;
import backtype.storm.Config;
import backtype.storm.LocalCluster;
import backtype.storm.topology.TopologyBuilder;
import backtype.storm.tuple.Fields;
import backtype.storm.utils.Utils;
public class WordCountTop {
    public static void main(String[] args) throws Exception {
        WordReader spout = new WordReader();
        WordNormalizer splitBolt = new WordNormalizer();
        WordCounter countBolt = new WordCounter();
        TopologyBuilder builder = new TopologyBuilder();
        builder.setSpout("word-reader", spout);
        builder.setBolt("word-normalizer", splitBolt).shuffleGrouping(
                "word-reader");
        builder.setBolt("word-counter", countBolt).fieldsGrouping(
                "word-normalizer", new Fields("word"));
        Config config = new Config();
        config.put("wordFile", args[1]);
        config.setDebug(false);
        config.put(Config.TOPOLOGY_MAX_SPOUT_PENDING, 1);
        LocalCluster cluster = new LocalCluster();
        cluster.submitTopology("word-count-top", config,
                builder.createTopology());
        Utils.sleep(10000);
        cluster.killTopology("word-count-top");
        cluster.shutdown();
    }
}
```

步骤 7

在 shanwei-starter/multilang/resources/ 目录下创建文件 word.txt，输入如下内容。

```
This is the first storm program!
This program is a common storm program!
Guangdong Shanwei Polytechnic
```

```
www.swvtc.cn
Welcome Shanwei
```

步骤 8

导出 jar 包，右击"swpt"，单击"Export…"，选择"Java"—"JVM file"，单击"Next>"打开 jar 包对话框，选择"swpt"，输入 JAR file 文件名，如图 11.23 所示，单击"Finish"完成后生产 jar 包。

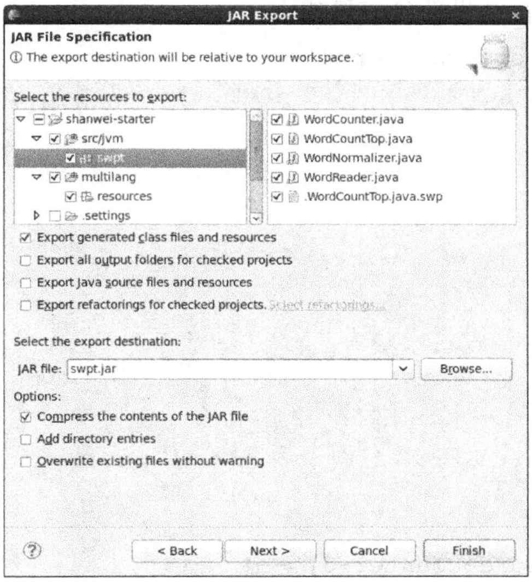

图 11.23　导出 swpt 包

步骤 9

运行程序，操作命令如下。

```
storm jar swpt.jar swpt.WordCountTop wcTest \
/root/workspace/shanwei-starter/multilang/resources/word.txt
```

运行后，在控制台显示如下部分信息如下。

```
-- Word Counter [word-counter-2] --
is: 2
www.swvtc.cn: 1
common: 1
a: 1
the: 1
storm: 2
Guangdong: 1
This: 2
Welcome: 1
program!: 2
program: 1
Shanwei: 2
```

```
Polytechnic: 1
first: 1
```
步骤 10

如果将 WordCountTop.java 中的下面代码改为

```
builder.setBolt("word-counter",countBolt,2).shuffeGrouping("word-normali
zer");
```

再重新编译运行，运行结果部分信息如下。

```
-- Word Counter [word-counter-2] --
is: 1
program!: 1
Shanwei: 2
a: 1
Guangdong: 1
storm: 1
This: 1
first: 1
Welcome: 1
……
-- Word Counter [word-counter-3] --
is: 1
program!: 1
www.swvtc.cn: 1
common: 1
program: 1
Polytechnic: 1
the: 1
storm: 1
This: 1
```

最后，请读者自己分析运行结果，并运行自己编写的拓扑实例。

11.5 本章小结

本章以任务为主线来说明 Storm 系统运行拓扑的情况。前两个任务以介绍为主，后两个任务突出实战。按照高职类学生学习能力和操作习惯进行编排，内容由浅入深，拓扑实例既简单而又有代表性，能很好地解释 Storm 系统的工作原理和机制。本章除了重点介绍 Eclipse 软件使用和打包的方法外，还介绍了如何使用 Maven 工具生成 Eclipse 项目包，并通过编写具体的拓扑实例——单词计数进一步加深读者对 Storm 工作原理的理解和对 StormUI 管理的熟悉程度。